The Physics Lab Manual
Part 1

Experiments to Accompany
Physics 1501 and Physics 2610 Laboratories
at Youngstown State University

Department of Physics and Astronomy
Youngstown State University

The Physics Lab Manual Part 1

Experiments to Accompany Physics 1501 and Physics 2610 Laboratories at Youngstown State University
Youngstown State University | Department of Physics and Astronomy | Fall 2019

Printed in the United States of America
10 9 8
ISBN: 978-1-61740-796-3

Van-Griner Learning
Cincinnati, Ohio
www.van-griner.com

President: Dreis Van Landuyt
Project Manager: Janelle Lange
Customer Care Lead: Lauren Houseworth

Clymer 796-3 Su19
312603-323708
Copyright © 2020

Acknowledgments

The YSU Department of Physics and Astronomy acknowledges the following for their assistance with this manual: Sharon Shanks, Physics and Astronomy, who designed and produced this manuscript; Carl Leet, YSU Media Center, who took the majority of the photographs, designed several of the diagrams and provided production support; Jerry Fullum and the crew of the YSU Engineering Shop, who helped design and constructed the apparatus for Experiment 4; Michael Schueller, B.S., 1994, who assembled much of the apparatus; Richard Pirko, Physics and Astronomy, who provided a valuable suggestion for the timing device used in Experiment 3; and Douglas Fowler and Ted Seman, Physics and Astronomy, for their many useful suggestions for improvement.

—William Cochran
Summer 1995

Additional acknowledgments to Sheila Bishop and Doug Fowler for help with proofreading and suggestions for improvements during the second revision.

—Jan Clymer
Fall 2003

Acknowledgments to Kathy Durrell for writing the math review added in the third revision, the drawing of the vernier caliper in the introductory pages and for the revision of instructions for the force probe in Experiment 6.

—Jan Clymer
Fall 2005

Additional acknowledgments to Greg Diamantis for updated pictures throughout the manual.

—Jan Clymer
Fall 2012

Table of Contents

Notes

The Physics Laboratory

READ ME

This manual contains instructions for the performance and analysis of experiments designed to supplement Physics lecture courses at Youngstown State University. The instructions are intended to be self-explanatory; normally no additional instructions will be given to you by your laboratory instructor.

Begin the experiment as soon as all members of your group are present. Large apparatus will have been assembled for you at your laboratory table. A tray containing smaller apparatus can be obtained from your instructor; be certain to obtain the tray that matches your table number.

ATTENTION!

Your success in the laboratory will depend largely on your preparation before beginning the experiment.

The instructions for making some measurements are intentionally vague, and you are expected to devise appropriate methods. Your instructor will assist you by answering questions and making suggestions where appropriate, but is present to observe your progress and evaluate your performance as much as to provide assistance.

Some of the experiments will not require the full scheduled time to complete, and you are free to leave when you have finished your measurements. Some of the experiments, on the other hand, may be difficult to complete in the time allowed. In these cases work rapidly and do as much as you can. *Your success in the laboratory will depend largely on your preparation before beginning the experiment.* Read the instructions before coming to the laboratory, and read the relevant sections of your lecture textbook.

Before writing your first report, it is *essential* that you read and understand the section **Precision and Accuracy** in this manual.

Notes

The Laboratory Report

An important objective for the student in the introductory physics laboratory is to learn the conventional methods for reporting experimental results. A laboratory report need not be long, but must be coherent and concise. Each instructor will have individual preferences for various parts of the report, but in general most reports follow the same form. In the absence of modifications or additions from your instructor, the following guidelines will produce an acceptable report. The guidelines have been modeled on the form of research articles in professional journals, which means you may assume you are directing your report to readers who will have as much knowledge of the subject as you have; lengthy descriptions of laboratory apparatus and well-known theoretical derivations normally need not be included.

The report consists of a title page followed by four sections: the introduction, the data, the calculations, and the analysis, as follows:

TITLE PAGE

In addition to the title of the experiment, the title page should include (a) your name, (b) the names of your laboratory partners, (c) your laboratory table number (stenciled on the apron of the table), (d) the date the experiment was performed, and (e) the date the report was submitted for grading.

INTRODUCTION

The introduction to your report should be brief and concise, written as if you were explaining the experiment to someone who has the same level of knowledge as your own. Theoretical derivations of equations are necessary only if your instructor or the laboratory manual calls for them, or in the uncommon event that you have based the experiment on an idea of your own with which others will not be familiar. Similarly, you need not spend time drawing diagrams of the equipment or explaining how a standard piece of laboratory equipment operates unless you have introduced modifications of your own design. For example, in Section 1, Experiment 2, you may state simply *"The Behr Free-Fall Apparatus was used to record the positions of a freely-falling plummet at 0.02 second intervals"*; a drawing is unnecessary since the apparatus is illustrated twice in the laboratory manual.

Only three things need to be included in the introduction:

1. A statement of **what** was *measured*,

2. An explanation of **how** the measurements were taken (and to what precision), and

3. A statement of the **purpose** of the measurements.

For example, suppose you were given a ruler and a block of wood approximately 100 cm long, 10 cm wide, and 1 cm thick and told to determine the volume of the block. An acceptable introduction would read, in part, something like this:

> *The length [**what**] of a block of wood was measured by means of a ruler [**how**] to a precision of one part in two thousand [see discussion below], the width to one part in two hundred, and the thickness to one part in twenty in order to determine the volume [**purpose**] of the block by means of the equation Volume = Length × Width × Thickness.*

No diagram needs be given of either the ruler or the block, and no derivation of the equation for the volume is required. Note that it is the length, width, and thickness that are *measured*, and the volume is *determined* from the measurements. The term precision as used in step (b) refers to the precision a given measuring device is capable of providing *relative to the size of the measurement being taken*. In the example it has been assumed that the ruler was divided into millimeter divisions and that estimates were made to the nearest half millimeter. Suppose then that this ruler is used to measure the length of the block, approximately 100 cm or 1000 mm, to the nearest 0.5 mm; 0.5 mm is 5/10000 or 1/2000 of 100 cm, so the precision is expressed as one part in 2000. Similarly, 0.5 mm is 5/1000 or 1/200 of 10 cm, so the relative precision of the width measurement is expressed as one part in 200, and for the thickness 0.5 mm is 5/100 or 1/20 of one cm and the relative precision is expressed as one part in 20. This method of specifying the precision reflects the fact that measuring a long length with such a ruler is more precise than measuring a short length; the ruler would provide a relatively precise determination of the length of a 10-meter laboratory bench, but would not be a very good device to use to measure the thickness of a fingernail.

ATTENTION!

Laboratory work is, by nature, cooperative; data are taken as a group effort. But it should be clear that submission of work done by another, use of data acquired by another without permission and acknowledgment, and fabrication and alteration of data to fit a "desired" result are dishonest acts and will result in a course grade of F and the submission of a report of academic dishonesty to the student's dean and to the Associate Vice President for Student Services.

DATA

Data should be recorded in ink on standard notebook paper during the course of the experiment. The data should be recorded in a coherent form as measurements are taken so that recopying for the laboratory report is unnecessary. Initial and date the data sheet. *The data sheet should be reviewed **and** signed by your instructor* before you leave the laboratory.

CALCULATIONS

Calculations should be presented in *tabulated* form; there are examples in the **Precision and Accuracy** section of this manual as well as in Section 1, Experiment 2.

ANALYSIS

The analysis section is the most important section of the laboratory report. It need not be lengthy, but should present a *quantitative* discussion of the results. The emphasis should be on the sources of the imprecision reported, i.e., on the quantitative extent to which your measurements were able to illustrate or verify the particular physical law being studied. Any questions asked in the laboratory instructions should also be answered in this section of the report. *Specific* sources of imprecision or error should be discussed *quantitatively*, and vague generalizations are to be avoided. Further discussion of these points is given in the **Precision and Accuracy** section of this manual.

Notes

Precision and Accuracy in Measurement: A Guideline

INTRODUCTION

Approximation is inherent in measurement, and no measured parameter can ever be known exactly. For this reason an experimental result is considered complete only when a quantitative expression is given of the precision of the result, a statement is made that delineates the sources of the imprecision, and an analysis is presented of the relative contribution of each source. These notes are intended as an abbreviated guide to the conventional methods used to report experimental results. It is expected that you will refer to the assigned text on data analysis for a more thorough treatment.

TERMINOLOGY

Terminology used in error analysis is sometimes ambiguous. Consider for example these definitions from the *Random House Dictionary of the English Language:*

> ER/ROR: The difference between the observed or approximately determined value and true value of a quantity.

> UN/CER/TAIN: Not clearly or precisely determined; indefinite.

According to the above definition of error, the term applies only to a comparison between the true value of a quantity and the experimentally determined value of the quantity. But usually there is no "true" or "known" value for a quantity being measured; determination of the value is the purpose of the measurement. In the physical sciences the term error, in fact, is used to mean two different things: error in the dictionary sense, and as an expression of the uncertainty, or imprecision, in an experimental result. The meaning of the term must be judged from context.

The sketches in Figure 1 illustrate the difference between the concepts of accuracy and precision. The four sketches represent marksmanship targets; the "true" or "correct" result for the marksman would be striking the bullseye, but this does not happen on every firing of the weapon. In Figure 1 (a) the marksman has obtained a precise result, for all firings have produced a closely grouped cluster of marks (i.e., the uncertainty is small), but the accuracy is poor (i.e., the error is high) since the cluster is not centered near the bullseye. Figure 1 (b) represents both low precision (large uncertainty) and poor accuracy (high error), and Figure 1 (c) high precision and high accuracy. What about Figure 1 (d)?

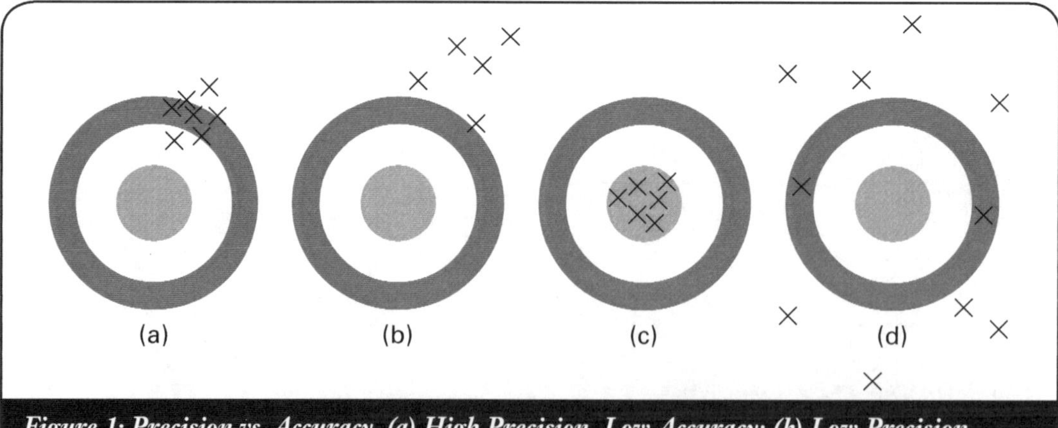

Figure 1: Precision vs. Accuracy. (a) High Precision, Low Accuracy; (b) Low Precision, Low Accuracy; (c) High Precision, High Accuracy; (d) ?

AN EXAMPLE

In the early part of this century physicists were studying the character of light emitted by a heated object. Metal, for example, glows red when heated; heated further the color changes, and when a sufficiently high temperature is reached the metal appears nearly white. Such emissions were studied carefully, the relative amounts of energy emitted at different frequencies (colors) being the measured quantity. A careless investigator might have graphically presented the result as shown in Figure 2.

At the time the measurements were being made, a theory had been developed to predict what the results should look like. Presented graphically, this theoretical prediction would look something as shown in Figure 3.

A comparison of Figures 2 and 3 will show an apparent agreement at smaller frequencies but an apparent disagreement at higher frequencies. But, in fact, it's not clear that there is any correlation between these two graphs because the investigator did not indicate how uncertain the experimental results were. Suppose instead the experimental graph had shown the uncertainty in the results by including with each point a small "error" bar. If the results had looked like Figure 4 (the theoretical plot has been superimposed on the data to make comparison easier), then there could be no doubt that a serious discrepancy exists between theory and experiment. This is in fact what happened. As a

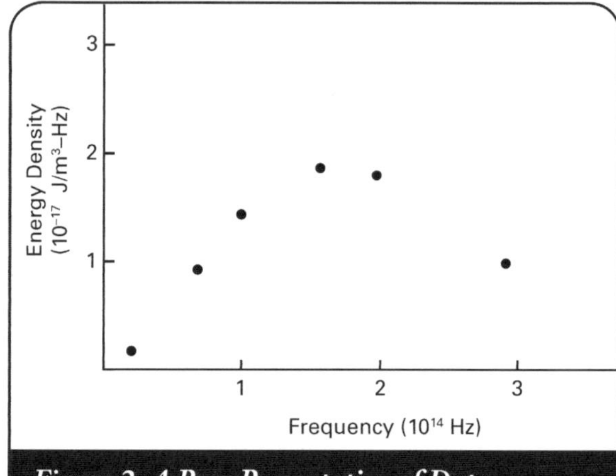

Figure 2: A Poor Presentation of Data

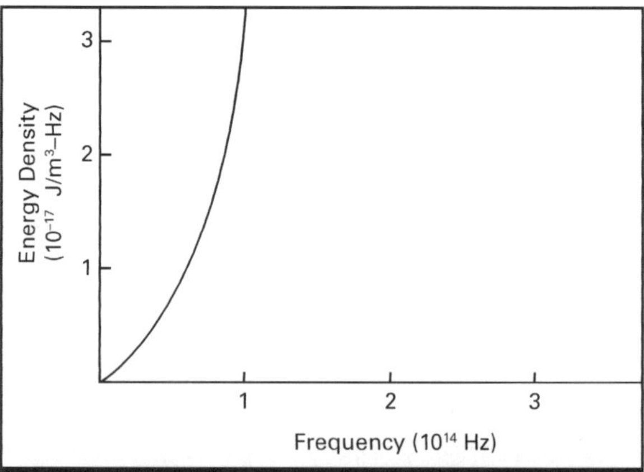

Figure 3: A Theorized Prediction of Results in Figure 2

result, a new look was taken at the theory, and the required modifications formed the basis of the quantum theory of radiation.

On the other hand, if the experimental results had looked like Figure 5, then the theoretical curve could still conceivably be consistent with the experimental results. What would then become crucial would be a knowledge of what physically caused the experimental uncertainties to be so large. The experiment could then be redesigned to reduce these uncertainties, and a more meaningful comparison of theory and experiment could be made.

THE MATHEMATICS
DISTRIBUTION IN RANDOM ERROR

When a parameter is to be measured the measurement is repeated many times in order to achieve an average value, the principle being that random fluctuations in the result are as likely to produce a value somewhat larger than the average as they are to produce a result somewhat smaller.[1] There are in general many possible causes of the fluctuations, such as estimation of scale readings by the observer, irregularities in the parameter being measured, etc. The fluctuations obtained can be illustrated by means of a *histogram*. For example, in Experiment 1 the time for the sand to empty from a sandglass timer is measured using an electric stop clock capable of recording time intervals to the nearest hundredth of a second. Repetitive measurements might produce results something like this:

Trial #1:   181.25 seconds
     2:     180.05
     3:     178.95
     4:     179.55
     5:     180.05
     6:     180.05
     .   .
     .   .
   100:     178.95
     .   .
Average: 180.00

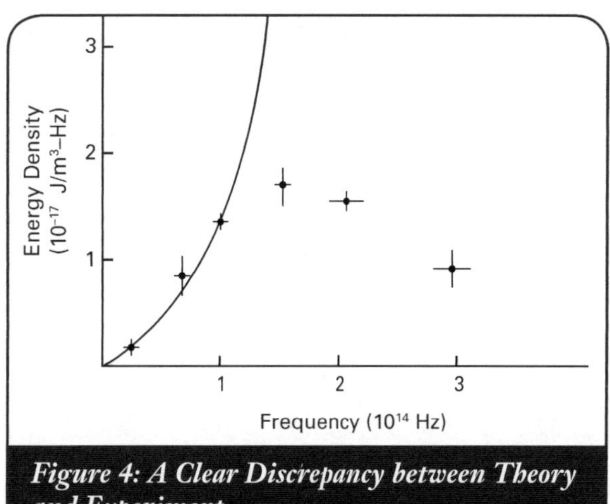

Figure 4: A Clear Discrepancy between Theory and Experiment

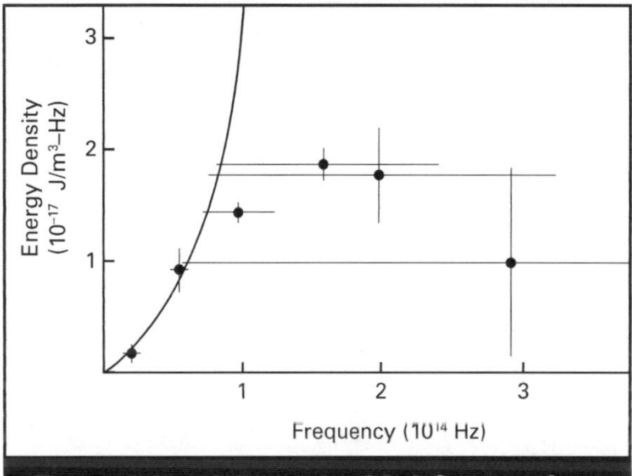

Figure 5: Result Requiring Modification of Experimental Procedure

___
Note

[1] See Section 1.4 in Lyons, Louis, *A Practical Guide to Data Analysis for Physical Science Students*, Cambridge: Cambridge University Press, 1991.

The histogram is constructed by drawing a vertical and a horizontal axis and dividing the horizontal axis into time second, centered about the average of the measured values. For each interval on the horizontal axis, the number of times a measured value was obtained which lay in that particular interval is plotted on the vertical axis. The result will look something like Figure 6.

If the distribution of values about the average is due to random causes, then if the size of the intervals on the horizontal axis is made smaller and smaller, the histogram will acquire a symmetric shape, and the points defining the centers of the tops of each vertical bar in the histogram will define a smooth curve, the *normal* or *Gaussian* distribution. This can be shown mathematically, and is known as the Central Limit Theorem. Such a curve is shown in Figure 7. The vertical axis must be reinterpreted as a representation of the *probability* that the corresponding value on the horizontal axis will be obtained on any given measurement, and the scale of the vertical axis is chosen so that the area under the curve has unit value.

The average value of the measured quantity is the numerical value on the horizontal axis directly under the peak of the distribution

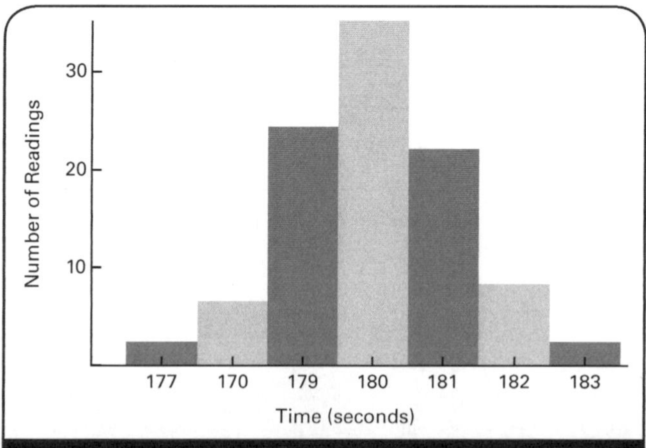

Figure 6: Histogram Showing History of Occurrence of Time Values out of a Total of 100 Readings

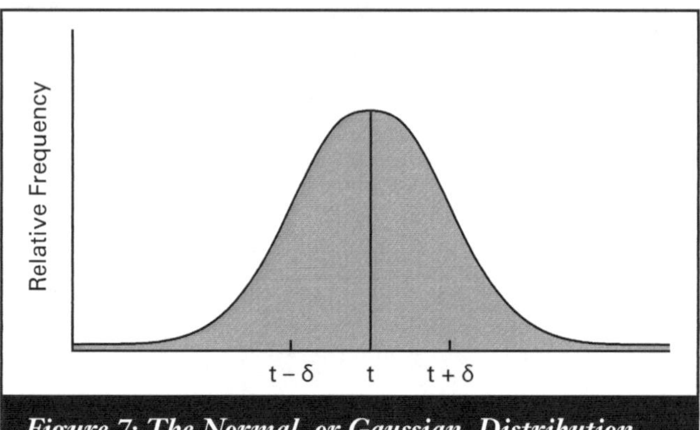

Figure 7: The Normal, or Gaussian, Distribution

curve, and clearly the width of the curve is related to how precisely the quantity has been determined. Convention dictates that the width of the curve is specified by the distance between the two points on either side of the average value which fall where the curvature of the distribution changes from concave to convex (some convention must be established to specify the width, since the curve never falls to the horizontal axis). It turns out that the width between the points where the curvature changes can be calculated from the table of experimental results. To the table is added a second column representing the difference (or deviation) between each experimental value and the average:

Trial #1:	t = 181.25 seconds	deviation = 1.25 seconds
2:	180.05	0.05
3:	178.95	−1.05
4:	179.55	−0.45
5:	180.05	0.05
6:	180.05	0.05
.	.	.
.	.	.
.	.	.
100:	178.95	−1.05
.	.	.

The standard (or rms) deviation, which is the distance on the horizontal axis of the Gaussian distribution curve from the average value to the change of curvature point on either side of the average (Figure 7) is found by squaring each value in the deviation column, adding the squares, dividing by the number of entries in the table,[2] and taking the square root of the result, i.e.,

$$\delta_t = \sqrt{\frac{\Sigma_i (t - \bar{\bar{t}})^2}{n}}$$

where t_i is the ith experimental value, $\bar{\bar{t}}$ the average value, and n the number of measurements taken. The experimental result is then reported as

$$t = \bar{\bar{t}} + \delta_t$$

In fact in many experimental measurements the above calculation is *not* justified: For the histogram of Figure 6 to transform into the Gaussian distribution of Figure 7 *many* measurements would have to be represented. How many? The answer is subjective, and the following guidelines may be used to determine how best to present a quantitative measure of the precision of an experimental result:

1. If more than 50 trials have been taken, represent the uncertainty by the rms deviation.

2. If more than 10 but fewer than 50 trials have been taken, report the uncertainty as the *average of the absolute values* of the quantities in the deviation column. (Why absolute values?)

3. If fewer than 10 measurements have been taken, report the uncertainty as the largest of the individual deviations.

4. If only one or two experimental values are recorded, the observer must make a reasonable estimate of the uncertainty, based on experience with the measuring apparatus and the characteristics of the object being measured.

PROPAGATION OF UNCERTAINTY

Often in experimental work the quantity that is measured is not the result sought, but rather is used along with other measured quantities to *calculate* the desired result. Since each measured quantity is uncertain by some amount there will be uncertainty in the calculated result as well. Propagation of uncertainty deals with how the uncertainty in the calculated result is determined from the uncertainties in the measured quantities. The correct procedures follow from calculus, but the following approximate methods will suffice for most calculations in this laboratory:

If a quantity x is to be calculated from two experimentally determined parameters y and z, with average results $\bar{y}$ and $\bar{z}$ and respective uncertainties δ_y and δ_z, then

    a. for addition or subtraction,
$$x = \bar{y} + \bar{z} \quad or \quad x = \bar{y} - \bar{z}$$
$$\delta_x = \delta_y + \delta_z$$

        (i.e., the uncertainty in x is the sum of the uncertainties in y and z)

Note
[2] See Section 1.4 in Lyons, Louis, *A Practical Guide to Data Analysis for Physical Science Students,* Cambridge: Cambridge University Press, 1991.

b. for multiplication or division,

$$\bar{x} = (\bar{y})\,(\bar{z}) \quad or \quad x = \frac{\bar{y}}{\bar{z}}\,,$$

$$\frac{\delta_x}{x} = \frac{\delta_y}{y} + \frac{\delta_z}{z}\,,$$

(i.e., the fractional uncertainty in x is the sum of the fractional uncertainties in y and z);*

c. for powers and roots,

$$x = \bar{y}^n,$$

$$\frac{\delta_x}{x} = n\!\left(\frac{\delta_y}{y}\right)$$

d. for logarithmic relations,

$$x = \log \bar{y},$$

$$\delta_x = (.434)\,\frac{\delta_y}{\bar{y}}$$

NOTE!

All uncertainties and fractional uncertainties should be taken as positive. The percent uncertainty for any quantity can be found by multiplying the fractional uncertainty by 100.

e. for trig functions,
If $x = \sin y$, $\delta x = \cos y\,\delta y$.
If $x = \cos y$, $\delta x = \sin y\,\delta y$.
If $x = \tan y$, $\delta x = \sec^2 y\,\delta y = \delta y\,/\,\cos^2 y$.

The five rules given above are approximations to results derived from calculus and are sufficient for the calculations required for this laboratory. A more complete discussion of error propagation can be found in *A Practical Guide to Data Analysis for Physical Science Students*, Cambridge University Press, 1991.

___________ *Note*
The factor of .434 arises from the fact that $\log y = (\log e)(\ln y)$, so $\delta(\log y) = (\log e)\,\delta(\ln y)$. Since $\log e \approx .434$, this gives $\delta(\log y) \approx (.434)\,\frac{\delta_y}{y}$

SIGNIFICANT FIGURES

When doing simple calculations with electronic calculators or computers there is often temptation to report all the digits registered by the calculator, but it should be clear that the reporting of more figures than are justified by the uncertainty in the specified quantity is meaningless. For example, if in the earlier example of the sandglass measurement the value obtained by averaging over all readings is 180.78 seconds, yet the resultant rms deviation turns out to be 2 seconds, it makes no sense to report the experimental result as 180.78 ± 2 seconds. If it is the first figure to the left of the decimal point that is uncertain, then there can be no significance to the two figures to the right of the decimal point. The correct expression of the result would be 181 ± 2 seconds.

When combining numbers to calculate a result, two guidelines should be followed:

    a. When adding or subtracting, keep no more decimals in the result than are in the parameter having the fewest decimals. For example,
$$26.135 + 0.31 = 26.45$$

    b. When multiplying or dividing, the result will have the same number of significant figures as the factor that has the smallest number of significant figures. For example,
$$(315.26) \times (66.8) = 211 \times 10^4$$

PERCENT ERROR

In rare instances there is a generally accepted value for the quantity that is to be determined experimentally. In such cases the error in the experimental result is reported as a percent error:

$$\% \ error = \frac{|accepted\ value - experimental\ value|}{accepted\ value} \times 100$$

Then the analyst must ask whether the accepted value, if not in agreement with the mean experimental value, lies within the reported range of uncertainty of the experimental value and if not, why not. An important point to keep in mind is that the result reported by the laboratory observer *is* the correct result for the experiment as performed under the given laboratory conditions. Errors are not made by carelessness or forgetting to take an important measurement; citing carelessness as a source of error is not acceptable. The source of error should rather be sought in a difference between the experimental conditions under which the observer worked and the conditions which existed when the "accepted" value was determined. Such conjectures should be considered *quantitatively*. For example, a piece of wire will have different lengths when measured at different temperatures; knowing the coefficient of thermal expansion of the wire, the temperature difference required to produce the reported error should be calculated and compared with the known temperature difference.

COMPARISONS

As stated above, only rarely is the laboratory observer provided with an accepted value with which the experimental result can be compared. Often, however, the same parameter will be determined by two or more different methods in an effort to improve precision and detect possible *systematic* errors. Then the different results are compared using a percent difference:

$$\% \ difference = \frac{|experimental \ value \ 1 - experimental \ value \ 2|}{average \ of \ experimental \ values \ 1 \ and \ 2} \times 100$$

In such a case an analysis of the relative precisions of the methods should be made; physically why did one method yield a result with more (or less) uncertainty than the other? Such considerations must be quantitative.

GOOD VS. POOR RESULTS

Can a result be characterized as good, poor, fair, etc.? The answer is that it is the use to which the results will be put that determines whether the measurement has been sufficiently precise. Knowing the speed of a runner to the nearest foot per second may be sufficient for predicting the time for a jog around the block, but insufficient for determining the result of a track meet. The only good analysis of an experiment is a quantitative analysis, and subjective generalizations should be avoided. Let the numbers speak for themselves.

Using a Vernier Caliper

A **vernier caliper** is a device that allows a precise measurement around an object (using the jaws) or of the space between two edges (using the arms). (A vernier can also be used for depth measurements, but we won't make use of that feature in this course.) See Figure 8.

To measure the length of an object, loosen the thumbscrew on the top of the vernier and insert the object between the jaws of the vernier. Then move the slide roller to adjust the gap between the jaws until it fits precisely the dimension you wish to measure. You will notice that there are two separate scales along the bottom of the vernier. (There are two scales along the top also, but these scales use British units, so ignore them.) There is a fixed scale that is marked in centimeters, with smaller markings to indicate tenths of a centimeter (i.e., mm). In addition, there is a sliding scale with eleven markings. This is the vernier scale. Take a reading from the fixed scale at the point where the leftmost mark of the vernier scale abuts it. If the mark on the vernier is between two marks on the fixed scale above it, use the smaller of the two. So, for instance, in Figure 9 on the next page, the leftmost mark on the vernier scale is between 1.1 and 1.2 on the fixed scale, so we use 1.1. (See A in the diagram.) Next examine the caliper carefully and determine which of the ten markings beyond the leftmost one on the vernier scale lines up exactly with a marking on the fixed scale above. This number gives the reading for the hundredths place for your measurement. In Figure 9, the sixth line on the vernier scale lines up exactly with the line on the fixed scale above it. (See B in the diagram.) So the value you would record for this measurement would be 1.16 cm.

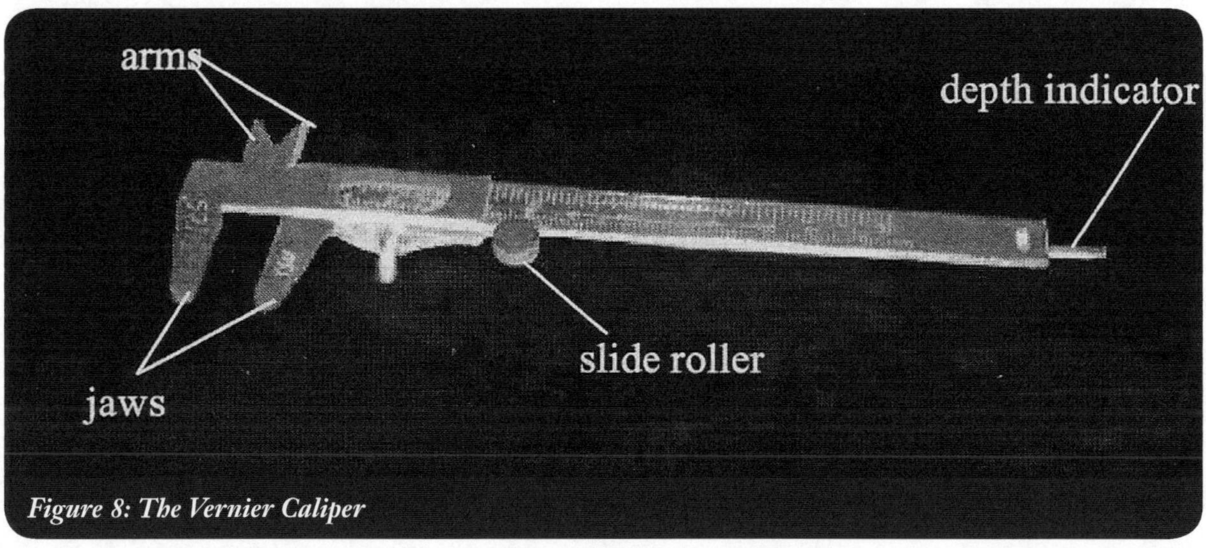

Figure 8: The Vernier Caliper

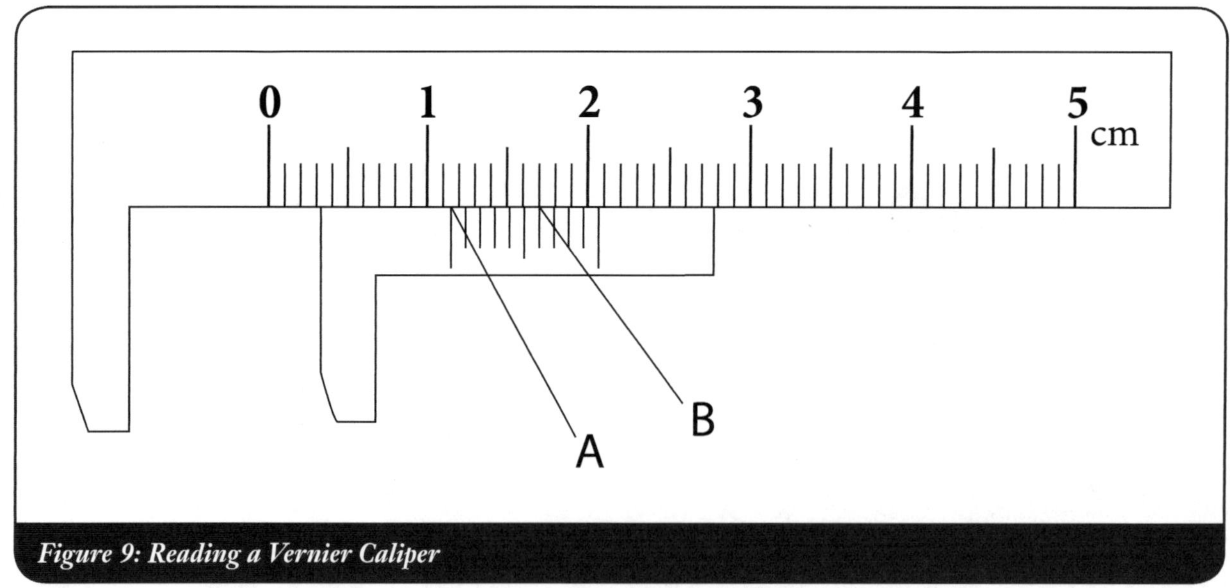

Figure 9: Reading a Vernier Caliper

For practice using a vernier caliper, go to *http://www.wcsscience.com/vernier/caliper.html*. (You will notice that the vernier scale there is labeled in mm. With a little thought, it should be evident that there is no difference between that and the cm label on the main scale on the photograph and the diagram above.)

Math Review

There are some basic math skills you will need in order to succeed in this lab. This section is intended to be a review of those skills. If any of the topics presented here are unfamiliar to you, you should consult a math text to get a more complete explanation of that topic.

BASIC ALGEBRA AND MANIPULATION OF EQUATIONS

If

$$a = \left(\frac{b}{c}\right)$$

then

$$b = ac \quad \text{and} \quad c = \frac{b}{a}$$

If we also know that $a = bd$, then by combining the two equations:

$$a = bd = \frac{b}{c}$$

and

$$d = \frac{1}{c} = c^{-1}$$

ADDING THE INVERSE OF TWO NUMBERS

$$a^{-1} + b^{-1} = \left(\frac{1}{a}\right) + \left(\frac{1}{b}\right)$$

$$= \left(\frac{b}{b}\right)\left(\frac{1}{a}\right) + \left(\frac{a}{a}\right)\left(\frac{1}{b}\right)$$

$$= \frac{(a + b)}{ab}$$

Note that this is **NOT** the same as $\dfrac{1}{(a + b)}$

POWERS

If a number (a) raised to a power is multiplied by the SAME number raised to another power, the exponents add. For example,

$$a^2 a^3 = a^{(2+3)} = a^5$$

Similarly if a number raised to a power is divided by the SAME number to another power the exponents are subtracted:

$$\frac{a^5}{a^2} = a^{(5-2)} = a^3$$

This is NOT THE CASE if the bases are different.
i.e.,

$$a^2 b^3 \neq (ab)^5$$

LOGARITHMS

The two most commonly used logarithms are the base 10 logarithm (usually expressed as log or $\log_{10}$) and the natural logarithm (expressed ln) which is base e, where e = 2.7182818.... Both the common log and the natural log behave in exactly the same way in any manipulations. The only difference between the two is the base. You can also create a log in any base you choose; however, only base 10 and base e logs will be used in this lab.

The common log is defined by

$$\log a = \mathrm{n} \iff 10^{\mathrm{n}} = a$$

i.e., the common log of a is the power to which 10 must be raised to obtain a. (The notation "log" is understood to mean $\log_{10}$, i.e., base 10 log.)

Similarly the natural (base e) log is defined by

$$\ln a = \mathrm{n} \iff \mathrm{e}^{\mathrm{n}} = a$$

meaning that the natural log of a is the power to which e (2.7182818…) must be raised to obtain a. (The "ln" notation is understood to mean $\log_e$, i.e., base e log.)

Taking a log is the inverse of raising the base to a power. So, $10^{\log(\mathrm{n})} = \mathrm{n}$ and, conversely, $\log(10^{\mathrm{n}}) = \mathrm{n}$.

The log of a PRODUCT of two numbers is equal to the SUM of the logs of the two numbers i.e.,

$$\log (ab) = \log (a) + \log (b)$$

Similarly the log of the QUOTIENT of two numbers is the DIFFERENCE of their logs i.e.,

$$\log \left(\frac{a}{b} \right) = \log (a) - \log (b)$$

Now since a^{n} is just a multiplied by itself n times,

$$\log (a^{\mathrm{n}}) = \mathrm{n} \log (a)$$

If we have a power-law equation, $y = a\,x^n$, we can express this in the form of the equation for a straight line by taking the log of both sides:

$$\log y = \log (ax^n)$$

$$\log y = \log a + \log x^n$$

$$\log y = \log a + n \log x$$

$$\text{or} \quad \log y = n \log x + \log a$$

which has the same form as the equation of a straight line: $y = m\,x + b$ if we replace y by log(y) and x by log(x) as shown below. The slope of this line is then given by n and the y-intercept is given by log (a).

$$\log y = n \; \log x + \log a$$
$$\downarrow \quad \downarrow \quad \downarrow \quad \downarrow$$
$$y = m \quad x \quad + \quad b$$

This is a general result. Whenever y can be expressed as a constant multiplied by a power of x, the slope of the log-log graph for y vs. x will give the power of x, and the y-intercept will be the log of the constant.

(Note: The properties listed here apply to both common logs and natural logs, as well as to logs of any other base.)

TRIGONOMETRY

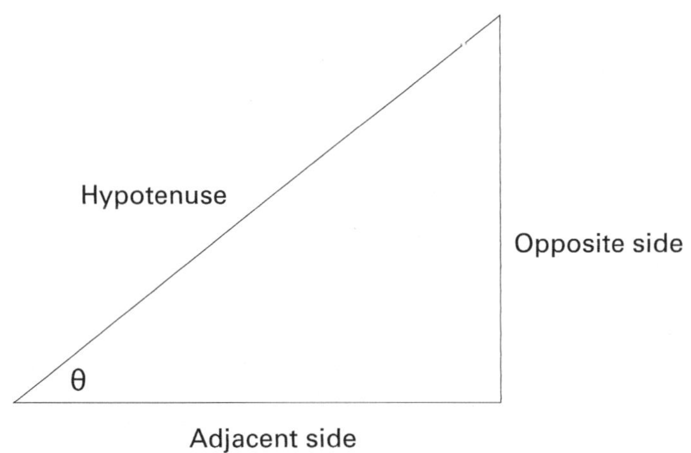

In a right triangle like the one shown above,

$$\sin \theta = \frac{opposite\ side}{hypotenuse} \; , \qquad \cos \theta = \frac{adjacent\ side}{hypotenuse} \; , \qquad \tan \theta = \frac{opposite\ side}{adjacent\ side}$$

Combining the above equations, it is easy to obtain,

$$\tan \theta = \frac{\sin \theta}{\cos \theta}$$

Sine, cosine and tangent are *periodic* functions, which means that they repeat in a regular pattern as the value of the argument (θ) changes. The sine and cosine functions are well behaved at all values of θ but the tangent function becomes undefined at values of $\theta = 90 \pm n\,(180°)$, where n is a positive integer. This means that the functions sine and cosine can be used to describe regular sinusoidal oscillations. The equation for these oscillations is given by:

$$y = A \cos (\omega t + \varphi)$$

where A is the amplitude of the oscillation, ω is the frequency of the oscillation and φ is the phase angle. Note that since $\sin (\theta + 90°) = \cos \theta$, either sine or cosine can be used in this equation and only the phase angle will change.

Below is a short list of trigonometric identities that will be useful for this lab. For a more complete list of identities (and the corresponding proofs), consult any of various mathematics texts.

$$\sin^2 \theta + \cos^2 \theta = 1$$

$$\sin (a + b) = \sin (a) \cos (b) + \cos (a) \sin (b)$$

$$\cos (a + b) = \cos (a) \cos (b) - \sin (a) \sin (b)$$

GRAPHING

The equation of a straight line is given by: $y = mx + b$ where m is the slope of the line and b is the y-intercept of the line. The slope of the line indicates the rate of increase of y with respect to the rate of increase of x. The slope is found as follows:

$$m = \frac{y_2 - y_1}{x_2 - x_1}$$

A slope of 1 indicates that x and y are changing at the same rate. The greater the slope, the faster y increases with respect to x. A line parallel to the x-axis has a slope of zero ($y_2 - y_1 = 0$), while one parallel to the y-axis has an infinite slope ($x_2 - x_1 = 0$). A positive slope indicates that the values of y increase as the values of x increase; a negative slope indicates that the values of y decrease as the values of x increase.

If you use a computer program to plot your graph and calculate the slope you should do a quick "sanity check" of the slope it gives you (especially if you use Excel). Just do a quick estimate of the slope from the above formula to make sure that the value the computer calculated is reasonable.

All graphs must have a title and axis labels. Each axis should be labeled with the quantity being plotted and its units. Axes must always be single-valued (no value repeats itself on the scale) and the scale must be linear (unless you are plotting a log-log graph using logarithmic graph paper or a graphing program on the computer, and then it is obviously logarithmic.) The graph must also be a full page in order to facilitate reading the data from it.

If you have a power-law equation the resulting graph will obviously not be a straight line if you use a linear scale. If the graph is not linear **do not** fit a straight line to it. If it appears to be a power-law you can do a power-law fit to it. (See page 21.)

A graph should also have error bars on all of the data points in both dimensions. These give you a measure of the uncertainty in the values being plotted.

Below is an example of a graph, done properly using the plotting program **Graphical Analysis** which is available on the computers in rooms 2010 and 2016.

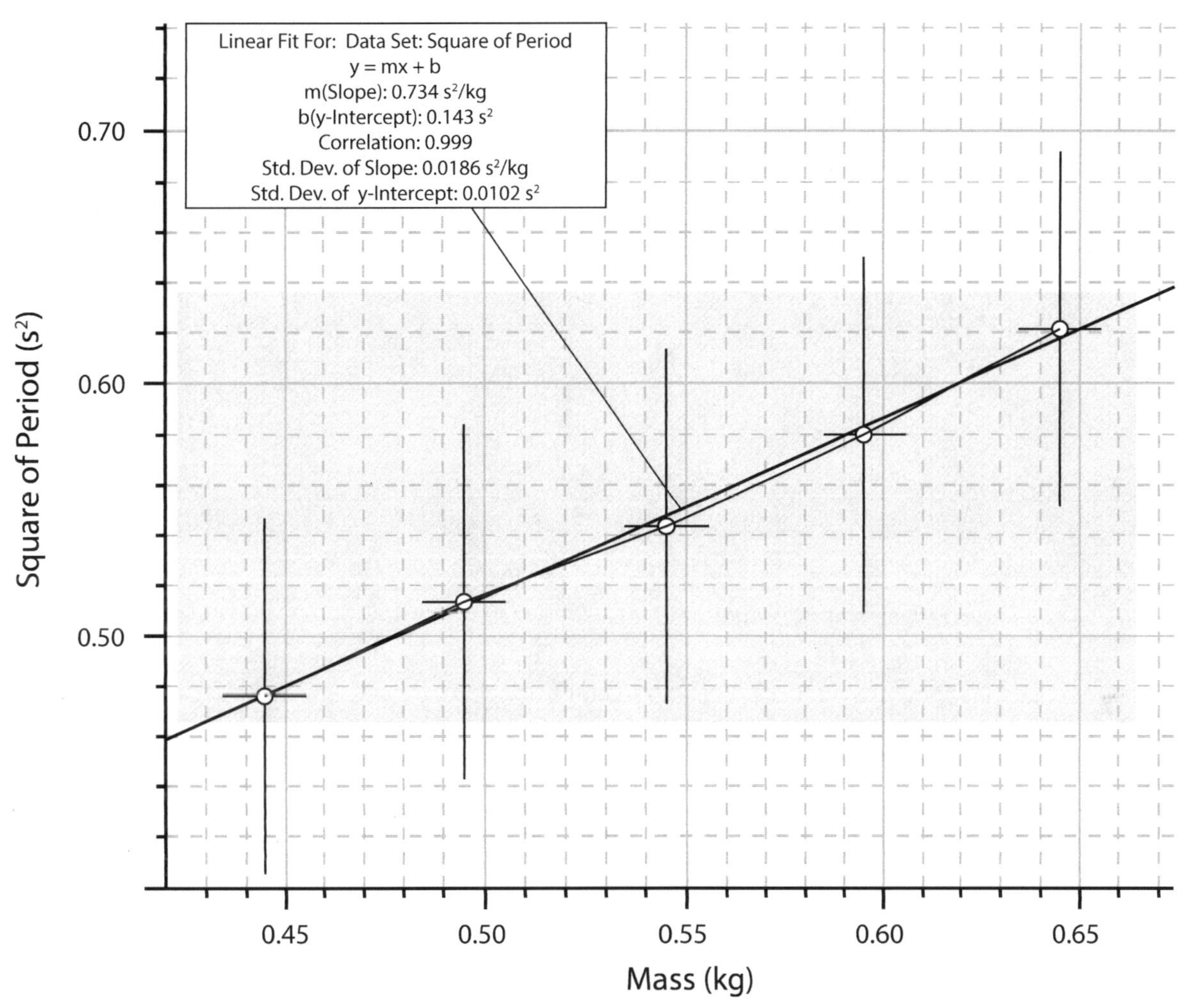

Notes

Summary of Important Tips on Graphing

1. On a y vs. x graph the independent variable (x) appears on the horizontal axis and the dependent variable (y) appears on the vertical axis.

2. All graphs should have a <u>title</u> and each axis should be labeled with <u>both</u> the <u>quantity</u> being graphed and the <u>units</u> in which it is measured.

3. Graphs should cover an <u>entire page</u>.

4. The slope should <u>always</u> be calculated from the graph, not from the data points.

5. If y is a linear function of x (that is, y depends on the first power of x), a least squares analysis can be used to obtain the best fit straight line. (**Excel** and **Graphical Analysis,** which are installed on the computers in Rooms 2010 and 2016, and also most other computer graphing programs include a least squares function.

6. If y is not a linear function of x (i.e., a graph of y vs. x does not produce a straight line), a power law should be tried. If y is a function of some power of x, a plot of log y vs. log x will produce a straight line, and the power can be obtained by the process described in the Logarithms section of the Math Review that precedes this section. (The graph can be obtained using **Excel, Graphical Analysis** or most other graphing programs, or by hand using log-log paper.)

7. If the dependence of y on x satisfies none of the above conditions, a smooth curve should be fitted to the data points. (**Graphical Analysis** will automatically "connect the dots" if you don't tell it to do anything else. This is not quite the desired smooth curve, but is the best option available in this case.)

Notes

EXPERIMENT 1

Measurement Precision and Distribution

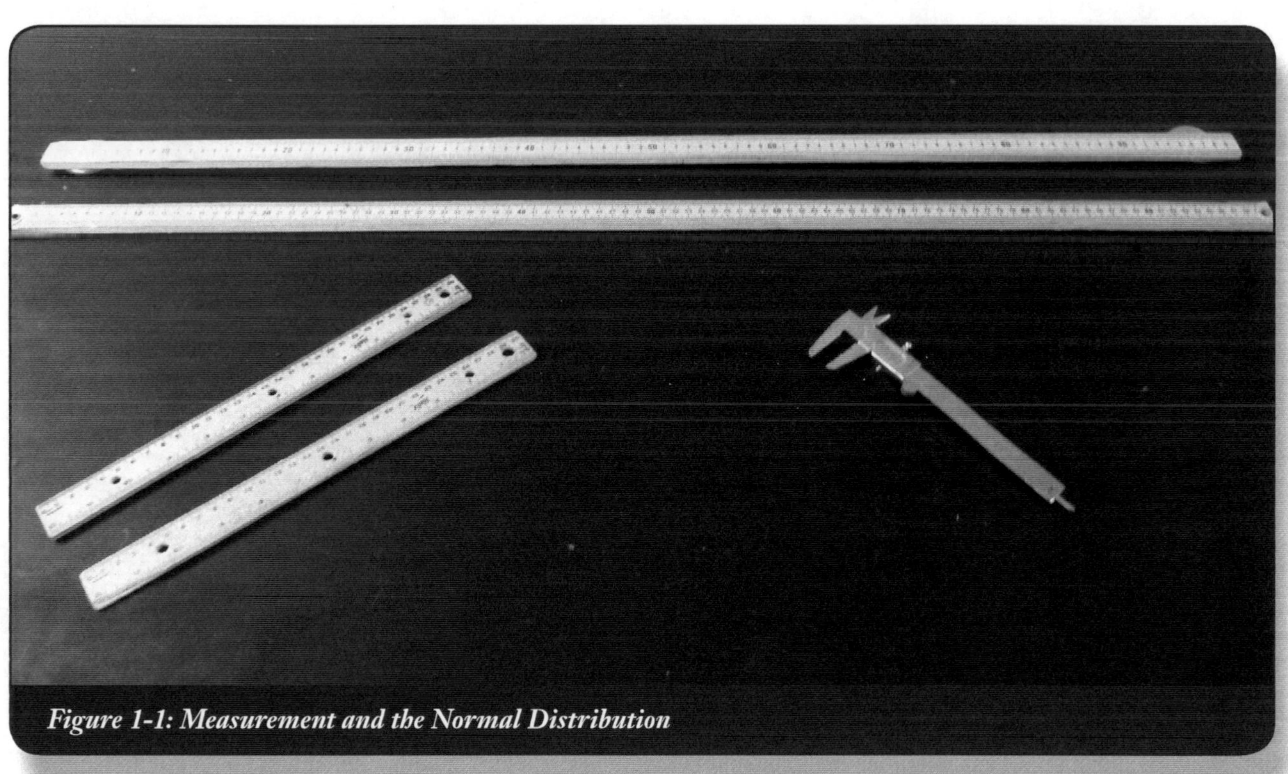

Figure 1-1: Measurement and the Normal Distribution

INTRODUCTION

The exercises in this laboratory session are designed to help you learn how to evaluate the uncertainty of your measurements in various situations and to determine how the uncertainty of those measurements combines to affect the uncertainty of results based on those measurements.

In the first exercise you will make one measurement of each of the dimensions of the top of your lab table and note the uncertainty of each of those measurements. You will then use those measurements and their associated uncertainties to calculate the volume of the table top and the uncertainty of that volume.

In the second exercise, you will make several measurements of the same quantity, i.e., the distance a ruler drops during the time it takes you to react and catch it. You determine the uncertainty of this measurement based on the distribution of the measured values about the mean for various numbers of trials. You will then use the distance measurement and its uncertainty to calculate the reaction time and its associated uncertainty.

The uncertainty of a measurement is best evaluated based on the distribution of many trials of measurement of the desired quantity about the mean. However, in some cases it may not be practical to perform a large number of trials. In some cases only one trial is available.

If only one measurement of a quantity is available, the measurement should be made as carefully and as precisely as possible, and the uncertainty should be evaluated based on the precision of the measuring instrument used. For example, if you are using a ruler on which the finest markings represent 0.1 cm, the uncertainty of the measurement would be reported as 0.05 cm. (You can tell which mark the dimension being measured is closest to, so the measurement shouldn't be off by more than half the distance between the finest marks, i.e., 1/2(0.1 cm) = 0.05 cm. Note that this method for determining uncertainty should be used **ONLY** when multiple trials are not available.

PROCEDURE

P1. Begin by using a meter stick to measure the length of the black top portion of your lab table. You will need to lay the meter stick down twice in succession in order to get a measurement of the total length of the table top. The uncertainty of each of these measurements can be determined as described above. Additionally, there is an uncertainty associated with the placement of the meter stick at the edge of the table or at the edge of the previous measurement each time you use the meter stick. These uncertainties should also be evaluated. Add all of these uncertainties to obtain the total uncertainty of the length measurement. Also add the two length measurements to obtain the total length of the table. Record the length of the table as the total length thus obtained plus or minus the total uncertainty. (Consult page 11 of your lab manual for determining the uncertainty of a sum of individual measurements.)

P2. Next use the ruler to measure the width of the table top. Evaluate the uncertainty of this measurement in the same manner as you did for the length in P1 and record the total value of the width plus or minus its uncertainty.

P3. Use the vernier caliper to measure the thickness of the table top. (If you don't know how to use the vernier, ask your instructor for help.) Evaluate the uncertainty in the same manner as for the length and width and record the value of the thickness with its associated uncertainty.

In the second part of the experiment, you will examine the effect of the number of trials on the uncertainty of a measurement.

P4. Begin by holding the end of the ruler between your lab partner's thumb and forefinger, so that the zero marking on the ruler is even with the top of your partner's fingers. Then drop the ruler, allowing your partner to catch it between the thumb and forefinger. (The partner catching the ruler should take care not to change the positions of the thumb and forefinger except to close them together and catch the ruler as it drops.) Read the measurement from the ruler at the position of the top of the fingers after the ruler has dropped and record this as the reaction distance.

P5. Repeat P4 for a total of 60 trials and record the results for each trial. (Don't forget to include units in your measurements.) Do not collect similar measurements together or count the number of times a particular value was obtained; simply make a list of the results of all 60 trials, in the order in which you measured them.

CALCULATIONS

C1. Check to make sure that the measurements obtained in P1–P3 for the length, width and thickness of the table top are all in the same units. Convert units if necessary.

C2. Calculate the volume of the table top by multiplying the length, width and thickness. Remember to keep the appropriate number of significant figures in your result. (If you aren't sure how to do this, review the rules for significant figures, which can generally be found in the first chapter of any physics text or on page 13 of your lab manual.)

C3. Calculate the fractional uncertainty of each of the measurements obtained in P1–P3 by dividing the uncertainty of each by the measurement.

C4. Now calculate the fractional uncertainty of the volume of the table top by adding the fractional uncertainties of the individual measurements for length, width and thickness. (See page 12 of your lab manual.) Again remember to keep the appropriate number of significant figures in the result.

C5. Calculate the uncertainty of the volume by multiplying the fractional uncertainty obtained in C4 by the volume obtained in C2. Record the final value of the volume plus or minus its uncertainty in the appropriate number of significant figures and with appropriate units.

C6. Calculate the percent uncertainty in your volume measurement by multiplying the fractional uncertainty by 100.

C7. Compute the percent difference between your value for the volume and that of the group next to you by taking the difference between the two values divided by the average of the two values and multiplying by 100. (See page 14 of your lab manual.) Check to see if the sum of the percent uncertainties obtained by each of the groups accounts for this difference.

C8. Ask your instructor to give you an "accepted" value for the volume of the table top. Compute the percent error for your measurement as compared to this accepted value by dividing the difference between the two values by the accepted value and multiplying by 100. (See page 13 of your lab manual.) Does the percent uncertainty you computed in C6 account for this error?

C9. Use the dimensions obtained in P1–P3 to calculate the perimeter of the tabletop and its uncertainty. Remember that since dimensions are being added here, it is the uncertainties, not the fractional uncertainties of the dimensions that must be added in order to get the uncertainty of the perimeter. Also calculate the percent uncertainty of the perimeter. Compute the percent difference between your value and that of the group next to you and the percent error, compared to the "accepted" value given by your instructor as before. Does your uncertainty account for the difference and the error?

Notice that there are actually three different ways to calculate the perimeter of the tabletop: one using the length and width, one using the length and thickness, and one using the width and thickness. Complete the calculations described above for at least two of these.

C10. Make a histogram showing only the first five values obtained in P4 and P5 for the reaction distance. Also calculate the mean value of the reaction distance for the first five trials.

C11. Now calculate the deviation of each of the individual values from the mean and record these results in a table with appropriate labeling and units. (The deviation for each value is obtained by subtracting the mean from each individual value of the measurement.)

According to your lab manual (page 11), when such a small number of trials of a measurement is available, the uncertainty should be recorded as the largest of the deviations of the individual values from the mean. Identify that value and record it as the uncertainty of the measured reaction distance in this case.

C12. Now use the first 20 of the measurements obtained in P4 and P5 and again make a histogram showing the distribution of the values. Calculate the mean and the deviation from the mean for each individual value. Record these results in a table as before.

If the number of trials for a measurement is between 10 and 50 as it is here, it is appropriate to report the uncertainty of the measurement as the average of the absolute values of the deviations. (Again see page 11 of your lab manual.) Compute this average and record it as the uncertainty in this case.

C13. Finally use all 60 of the measured values obtained in P4 and P5 to construct a histogram. Do you notice that the histogram is more closely approximating a bell curve as the number of trials increases?

C14. In this case, your lab manual tells you that the uncertainty should be reported as the rms standard deviation of the values from the mean. In order to compute this value, first calculate the mean value for all 60 trials and find the deviation of each value from the mean. Record these values in a table as before. Add another column in your table in which you record the square of each of the deviation values. Then compute the mean for the 60 values of the deviation squared. Finally, take the square root of that mean value. The value thus obtained is called the rms standard deviation. Record this as the uncertainty of the measurement of reaction distance for this case.

C15. Use the values obtained in C14 to compare the histogram you constructed in C13 to a Gaussian (bell) curve. For a Gaussian curve, 68% of the values fall within one standard deviation of the mean, 95.5% of the values fall within two standard deviations, and 99.7% fall within three standard deviations.

C16. According to principles that will be developed later in the course, the reaction distance, d, is related to the reaction time, t, by the equation

$d = 1/2\ gt^2$. This can be written alternatively as $t = (2d/g)^{1/2}$.

This equation can be used to calculate the reaction time if the reaction distance is known. We can also use it to calculate the uncertainty of the reaction time from the uncertainty of the reaction distance. According to the rules for propagation of uncertainty (See page 12 of the lab manual), the fractional uncertainty of t should be computed as ½ the fractional uncertainty of d. Compute t, and the fractional uncertainty of t, then multiply these to obtain the uncertainty of t. Record the calculated value of t with its associated uncertainty. Also calculate the percent uncertainty of t.

For further review of the concepts introduced in this experiment, review pages 7–14 of your lab manual and use the online tutorials (accessible from the physics department website) for further help if necessary.

Note

The exercises just completed should be viewed as examples of the type of analysis used to evaluate uncertainty in an experiment, not as a template for all later calculations. Each experiment presents slightly different challenges, and the evaluation of the uncertainty of a particular measurement will need to be tailored to the specific situation.

Notes

Free Fall and the Acceleration Due to Gravity

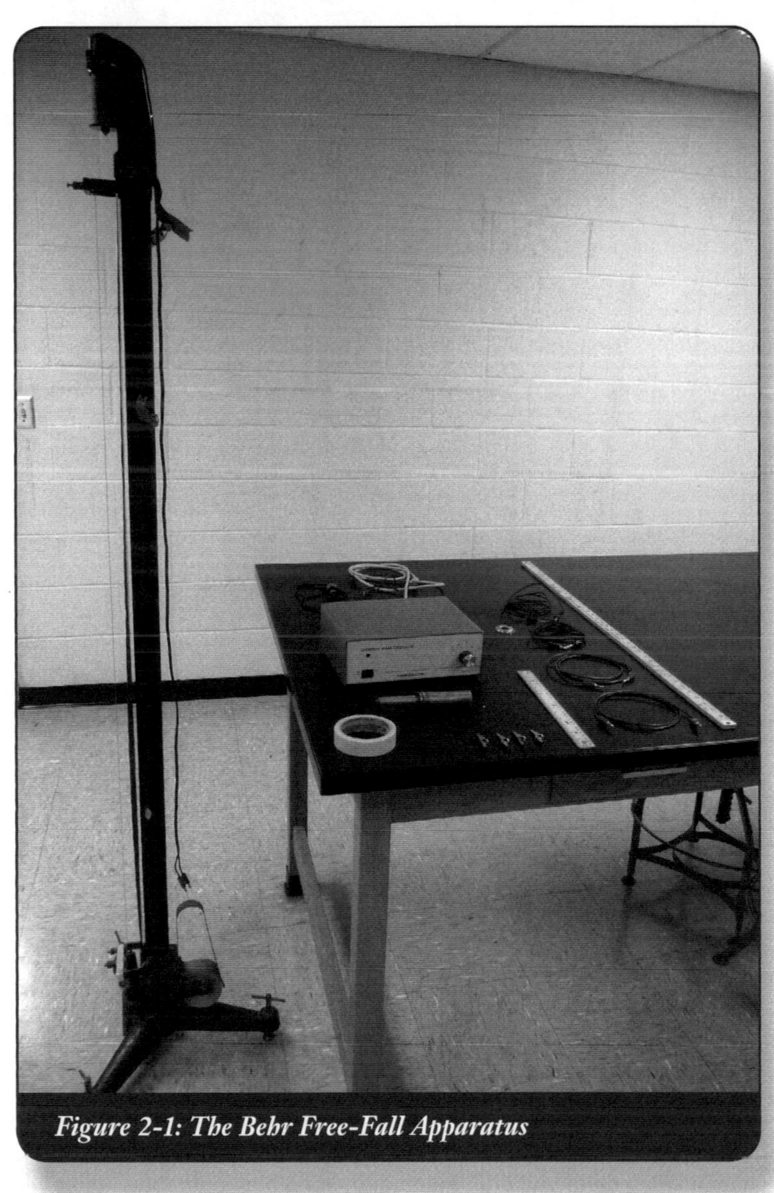

Figure 2-1: The Behr Free-Fall Apparatus

INTRODUCTION

The object in this experiment is to determine the acceleration due to gravity by measuring the positions of a freely-falling body at the ends of successive equal time intervals. In the Behr Free-Fall Apparatus (Figures 2-1 and 2-5), two wires about six feet long are stretched vertically about three inches apart. A plummet (Figure 2-2), consisting of a solid cylinder fixed with a metal collar, falls between the two wires while high voltage is applied periodically to the wires. Since the distance between each wire and the collar of the plummet is small, 1/8 inch or less, the high potential difference generates a spark between each wire and the collar. As the spark jumps from the collar to the rear wire, it passes through a thermally-sensitive tape, leaving a mark as the plummet passes. The series of marks thus provides a record of the positions of the falling plummet at known instants of time, the time interval between marks being determined by the spark-generating source.

Your instructor will demonstrate the operation of the apparatus and provide a tape for each group that will be used for your measurements. It is important that you pay attention during this demonstration and understand the principles involved.

PROCEDURE

P1. In addition to the Behr Free-Fall Apparatus, a high-voltage spark generator (Figure 2-3) and the low-voltage power supply mounted on the apron at the end of the laboratory table (Figure 2-4) will be used.

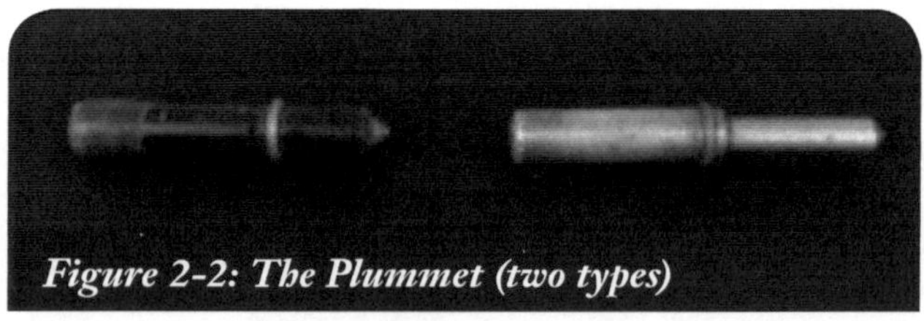

Figure 2-2: The Plummet (two types)

Figure 2-3: Type of Spark Source

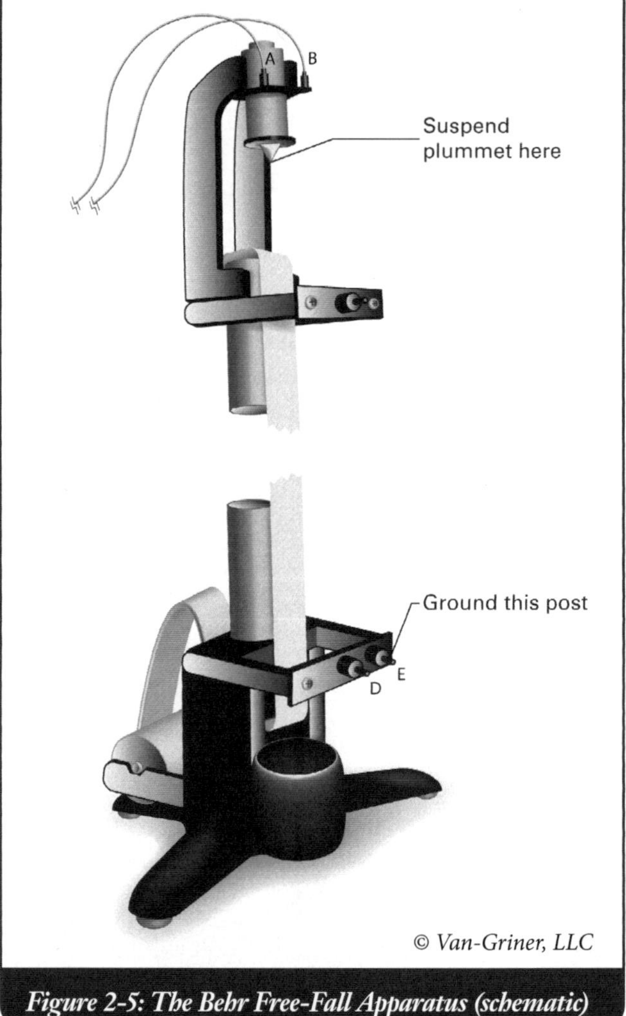

Figure 2-5: The Behr Free-Fall Apparatus (schematic)

Figure 2-4: The Lab-Volt Power Supply

P2. Notice that the top of the free-fall apparatus contains a coil of wire, the ends of which can be connected by wires to a voltage source, thus producing an electromagnet. That is, when voltage is supplied to the electromagnet, current flows through the coil, and the magnet is on, but turning the voltage to the electromagnet off turns the magnet off.

The binding posts A and B of the electromagnet are connected to the lower right-hand terminals (marked "B 0-18 V, 10 A, DC") on the Lab-Volt Supply (Figure 2-5). The range switch is set to "Range B." With the rotary dial set at "0," the key is used to turn on the power supply. Now, while holding the pointed end of the plummet against the pointed end at the bottom of the electromagnetic coil, the voltage is slowly increased by turning the rotary dial until the plummet is held in place by the electromagnet. A setting of about "2" on the rotary dial should be sufficient. Turning off the Lab-Volt supply will release the plummet when data is to be taken.

WARNING!

SHOCK HAZARD!

The spark source generates sufficient energy in each discharge to cause extreme discomfort. While the current produced is not normally dangerous to a healthy person, individuals vary in their tolerances to electrical shock, and care must be taken to avoid accidental contact with the high voltage output.

Make certain that you are not in contact with the spark generator or any part of the Behr Free-Fall apparatus itself, when the unit is on. Remember that sparks may be produced even if the remote starter is not being activated.

P3. Next the high-voltage output wire (red) from the spark generator is connected to terminal D of the free-fall apparatus (the one connected to the wire) and the ground wire (black) from the spark generator is connected to terminal E (the one connected to the base). The mode switch is then set to the "line" position. This provides a spark discharge at the line frequency of 60 Hz (i.e., one spark every 1/60 second). The spark should not discharge until the remote hand-held button is depressed, but be careful not to touch the free-fall apparatus while the spark source is on, as the spark source has been known to malfunction, causing a discharge without depressing the button on the remote control.

P4. When all necessary connections have been made, the spark timer will be turned on, using the remote switch and the Lab-Volt power supply will be turned off, thus turning off the electro magnet and releasing the plummet. As the plummet falls, the discharging spark passes through the thermally-sensitive paper. As soon as the plummet reaches the end of its drop, the spark generator is turned off. **Do not touch the apparatus while the spark source is on.**

P5. Your instructor will then remove the recorded tape from the apparatus. This process will be repeated to make a tape for each group to use. (Or a previously made tape may be given to each group after the operation of the apparatus has been demonstrated.)

Some things to watch for:

a. The plummet should fall smoothly, landing in the pocket at the bottom of the apparatus. If it doesn't, the apparatus may require leveling.

b. Care should be taken to ensure that the plummet does not swing as it falls; release the plummet only after it has stopped any slight swinging motion.

c. Just as a lightning bolt does not travel in a straight line, a small spark will wander in its path. See if you can detect evidence of this "wander," and estimate its effect quantitatively.

P6. Examine your tape carefully to make sure the spark marks are clear and distinct and there are no scratches or other marks that obscure the data. Your tape should contain at least 20 clearly-distinguishable marks that get gradually farther apart. Once you have obtained a good usable tape, use masking tape to secure it to the lab table.

P7. Starting from the end where the marks are closest together, use the meter stick or ruler to measure the distance Δy between each adjacent pair of marks (Figure 2-6). Record these distances for as many marks as you can clearly identify. Also record the uncertainty of your measurements. Stop taking measurements when the marks become unclear or they begin to get closer together rather than farther apart. (This indicates that the plummet hit the bottom and bounced upward.)

CALCULATIONS

C1. Calculate the average velocity for each time interval between marks by dividing each of the distances Δy measured in P7 by Δt for the interval, which is 1/60 second. (The time interval Δt between each pair of marks is 1/60 second, so the average velocity for each interval is the distance traveled between marks divided by that time. Also, notice that dividing by 1/60 is the same as multiplying by 60, so the average velocity for each interval can be found by simply multiplying each value of Δy by 60.) Use the uncertainty you recorded in P7 and a reasonable estimate of the uncertainty in Δt to determine the uncertainty of the velocity.

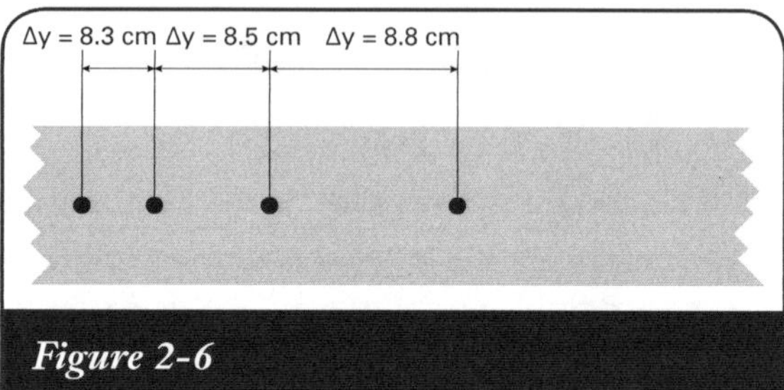

Figure 2-6

C2. Because the acceleration of the plummet is uniform (i.e., constant), the average velocity between any two marks is a good approximation of the instantaneous velocity at the time midpoint between the marks. For example, the instantaneous velocity at point a in Figure 2-7 is equal to the average velocity

$$\left(v = \frac{\Delta y}{\Delta t} \right)$$

between points A and B, the instantaneous velocity at point b is the average velocity between points B and C, etc.

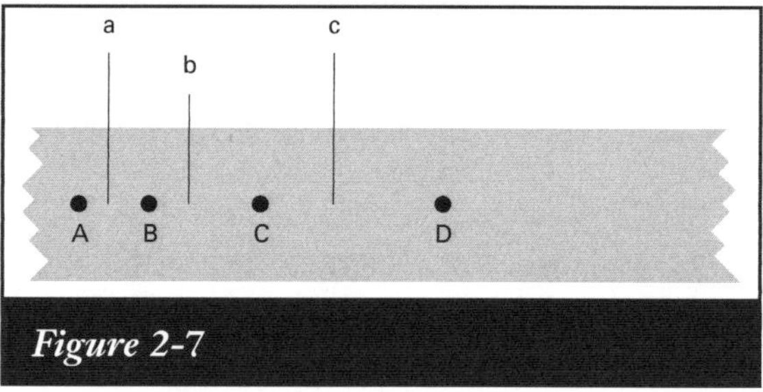

Figure 2-7

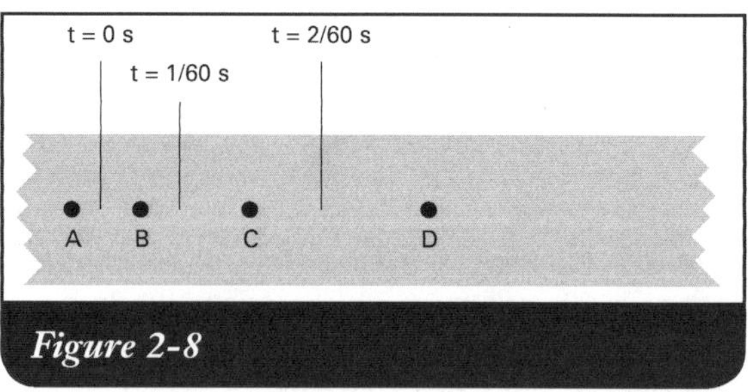

Figure 2-8

Note that the time midpoint is not the same as the displacement midpoint. Since the velocity of the plummet is constantly increasing, the displacement during the first half of any time interval will always be less than the displacement during the second half of the interval. Thus the time midpoint is shifted to the left of the displacement midpoint, as shown in Figure 2-8.

In order to track the change in velocity with time, take the zero of your running time to be at the time midpoint between the first and second marks (point a in Figure 2-7). The time midpoint between the second and third marks (point b) is then 1/60s, the time midpoint between the third and fourth marks (point c) is 2/60s, etc. (See Figure 2-8.) Make a table of the instantaneous velocity (from C1) vs. the running time. You will need to convert the running times of 1/60s, 2/60s, etc. to decimal equivalents and you may wish to convert the velocity values cm/s to m/s. (Notice that this running time is different from the Δt used for the calculations in C1, which was the same for each interval.) Table 2-1 shows a sample of how the first several entries in your table might look.

C3. Use Graphical Analysis or Excel or another graphing program of your choice to plot a graph of velocity vs. running time. Use the uncertainties from C1 to place horizontal and vertical error bars on your graph. Figure 2-9 shows an example using Graphical Analysis.

C4. Next use the computer to construct a linear fit for your data points and calculate the slope. (See Figure 2-10 for an example using Graphical Analysis.) The slope of this line will be your experimental value of the acceleration due to gravity. Calculate % error for your value compared to the theoretical value of g = 9.80 m/s² (or 980 cm/s² if you left your velocity values in cm/s.) (The linear fit can be obtained using either Graphical Analysis or Excel or most any other graphing program. Ask your instructor for help if you need it. A description of the process used to obtain a linear fit to data and the slope and y intercept of the resulting line, along with their uncertainties can be found in Lyon's *Practical Guide*[3].)

C6. You may use the standard deviation of the slope as the uncertainty of your slope and therefore the uncertainty in your value of g. (According to page 11 of your lab manual, you should have at least 50 data points in order to calculate standard (rms) deviation, but when a graphing program is used to obtain a linear fit and determine the slope and/or y intercept of a graph, an exception will be made to this rule.)

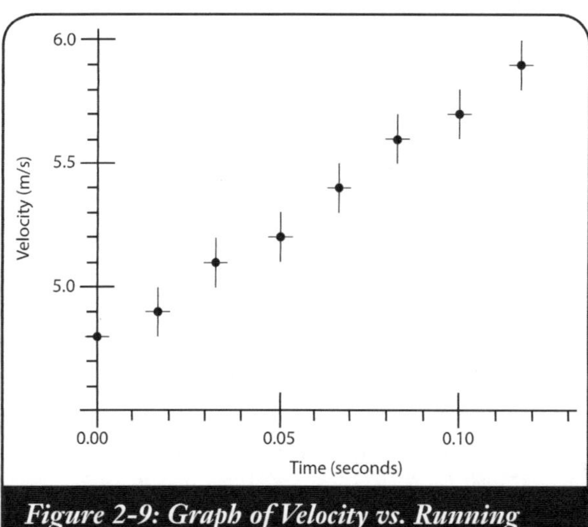

Figure 2-9: Graph of Velocity vs. Running Time (showing points and error bars)

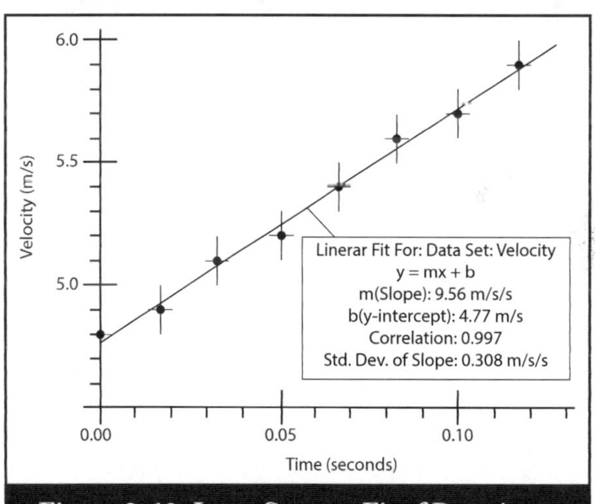

Figure 2-10: Least Squares Fit of Data in Table 2-1 (error bars not to scale)

C7. Based on your error and uncertainty calculations, can you conclude that your measured value of g is in agreement with the theoretical value? Discuss sources of error in this experiment and how they might impact your results.

Note

[3] Lyons, Louis, *A Practical Guide to Data Analysis for Physical Science Students*, Cambridge: Cambridge University Press, 1991.

Table 2-1: Displacement and Velocity vs. Time for an Object in Free-Fall

Running Time t (s)	Δy (cm)	V (cm/s)	V (m/s)
0	8.0	480	4.8
0.017	8.2	490	4.9
0.033	8.5	510	5.1
0.050	8.7	520	5.2
0.067	9.0	540	5.4
0.083	9.3	560	5.6
0.100	9.5	570	5.7
0.117	9.8	590	5.9

References for further study:

Yardley Beers, *Introduction to the Theory of Error*, Reading, Mass: Addison-Wesley, 1957.

Keith H. Burrell, "Error analysis for parameters determined in non-linear least-square fits," *Am. J. Phys* 58 (2), 160 (1990).

Sherril D. Christian, Edwin E. Tucker and Eric Enwall, "Least squares analysis: A primer," *Am. Lab* 18 (6), 41 (1968).

Lawrence Hmurcik, Amy Slacik, Hilda Miller and Sandy Samoncik, "Linear Regression analysis in a first physics laboratory,"*Am. J. Phys* 57 (2), 135 (1989).

B. Cameron Reed, "Linear least squares fits with errors in both parameters," *Am. J. Phys* 57, 642 (1989).

B. Cameron Reed, "Erratum for linear least-square fits with errors in both parameters," *Am. J. Phys* 58 (2), 189 (E) (1990).

Notes

Notes

Empirical Relations:
Free Fall

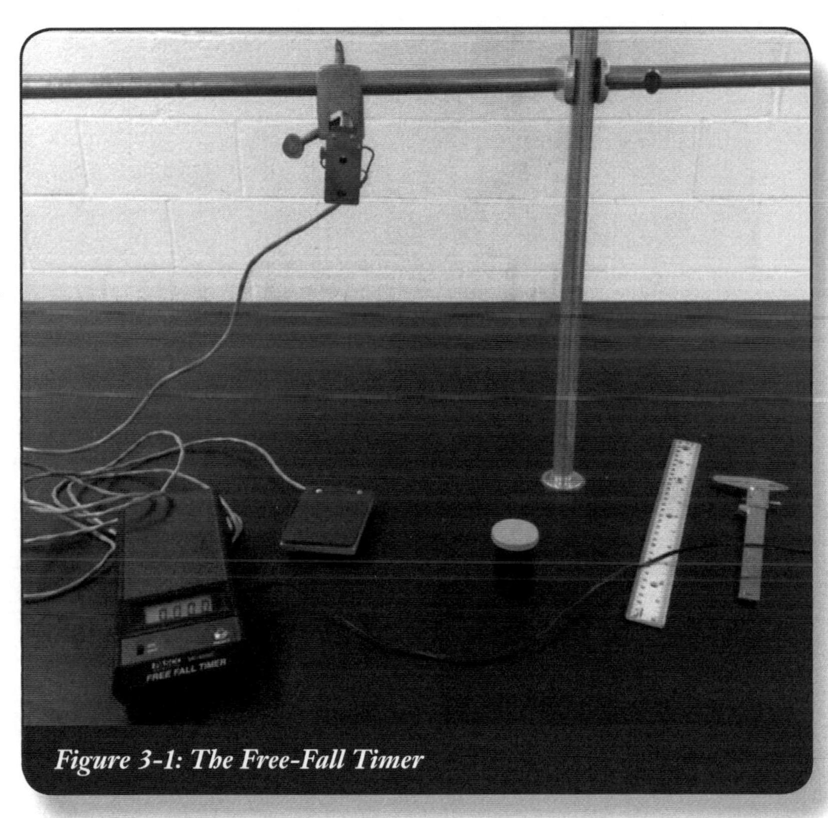

Figure 3-1: The Free-Fall Timer

INTRODUCTION

When a relationship is suspected to exist between two or more physical parameters, the relationship is sought in one of two ways: *theoretical* or *empirical*. In Experiment 2 the theoretical relations between distance, velocity and acceleration were assumed valid and the acceleration due to gravity was determined by substituting into these relations numerical values of distance and time obtained from experiment. In the present experiment a different point of view is taken: it is observed that the position of a falling object depends on the duration of its fall; this observation can be expressed mathematically as $y = f(t)$, i.e., vertical distance (y) is a function of time (t); it will be the task of *observation* to determine the exact form of this relationship.

The Free-Fall Timer used in this experiment provides a record of the time required for a steel ball bearing to fall a measured distance. To determine if the relation between y and t is linear, a graph is made, plotting y on the ordinate (y-axis) and t on the abscissa (x-axis). If the points appear to fall in a straight line, the method of least squares for a linear fit can be applied to find the slope and the ordinate-intercept of the graph and hence the numerical parameters needed to write the linear equation which relates y to t. When this procedure is followed in this experiment, however, it will not define a straight line, but rather curve upward, something like as shown in Figure 3-2.

The upward curve of the points suggests that the form of the equation relating y and t may be of the power law type, i.e.,

$$y = kt^n \qquad \textbf{(1)}$$

with $n > 1$ (a downward curvature would suggest $n < 1$; why?). To determine whether this is in fact the correct form, note that taking the logarithm of each side of Equation (1) yields:

$$\log y = n \log t + \log k \qquad \textbf{(2)}$$

If log y rather than y is now taken to be the ordinate variable, and log t as the abscissa, a plot of log y vs. log t will produce a straight line. If the logarithmic plot does, in fact, appear linear, then the relationship between y and t can be assumed to be of the power law form [Equation (1)], and a least squares fit can be made in order to obtain a more precise determination of the constants k and n and their relative uncertainties. (The slope of the log y vs. log t graph is n and the y intercept is log k.)

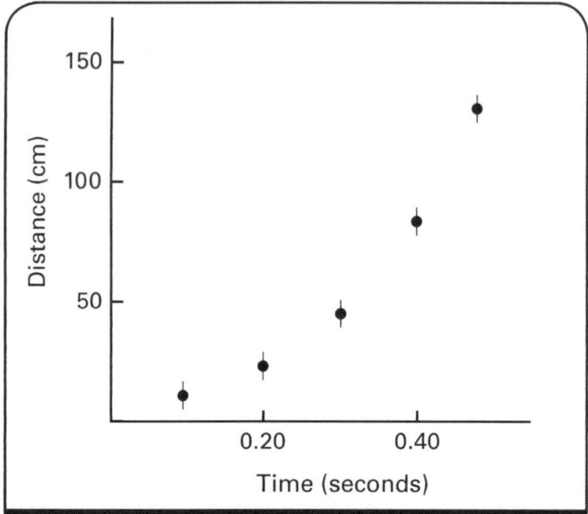

Figure 3-2: A Likely Experimental Result of Distance vs. Time for a Freely-Falling Object

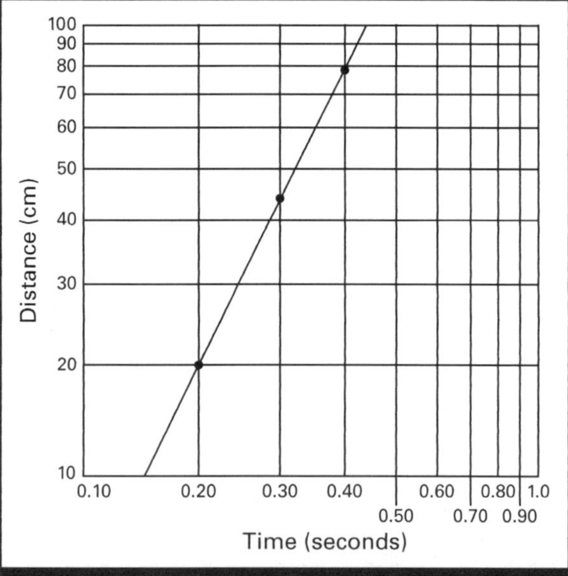

Figure 3-3: A Log-Log Plot of Distance vs. Time for a Freely-Falling Object

According to theory that can be found in any standard physics textbook, the relationship between distance fallen (y) and time elapsed (t) for a falling object starting from rest is given by

$$y = 1/2 \; g \; t^2 \qquad\qquad \textbf{(3)}$$

Comparing this to Equation (1), we see that the expected value of n is 2, and the expected value of k is 1/2 g. A percent error can be found for n and a value for g can be calculated and compared to the "book" value, and to the value you obtained in Experiment 2. Consideration can be given to which method, Experiment 2 or Experiment 3, gives the more precise determination and why.

PROCEDURE

P1. Examine the equipment carefully. In addition to two steel ball bearings of different size which serve as falling objects, there is an electronic timer and a target pad onto which the ball falls (Figure 3-4). Releasing the ball starts the timer; the impact of the ball on the target pad turns the timer off.

P2. There are two different types of timers. At some stations, a timer like the one shown in Figure 1 is used. This type of timer is permanently connected to the free-fall release mechanism and the pad. At other stations, a Smart Timer is used (Figure 3-5). In that case, the cord from the free-fall apparatus should be plugged into the terminal marked "1" on the right side of the Smart Timer. Both types of timer use a voltage adapter. Plug the cord attached to the adapter into the timer and plug the adapter into the socket on the lab table.

P3. With a steel ball placed in the release mechanism (Figure 3-4), use the plumb bob to position the target pad directly under the ball. The distance between the ball and pad will be varied during the experiment, and the ball-pad alignment should be checked each time the distance is changed.

P4. Place one of the steel balls in the release mechanism and turn the timer ON. (You may need to hold the ball in place until it is ready to be released.)

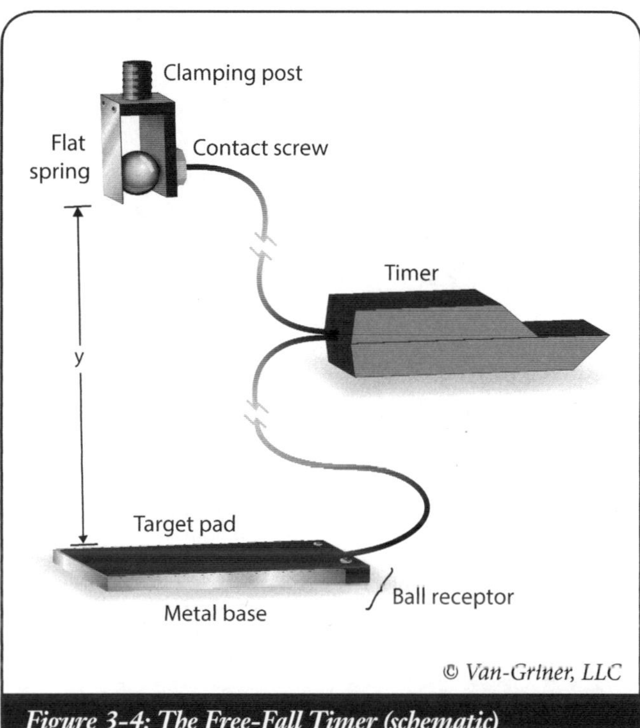

Figure 3-4: The Free-Fall Timer (schematic)

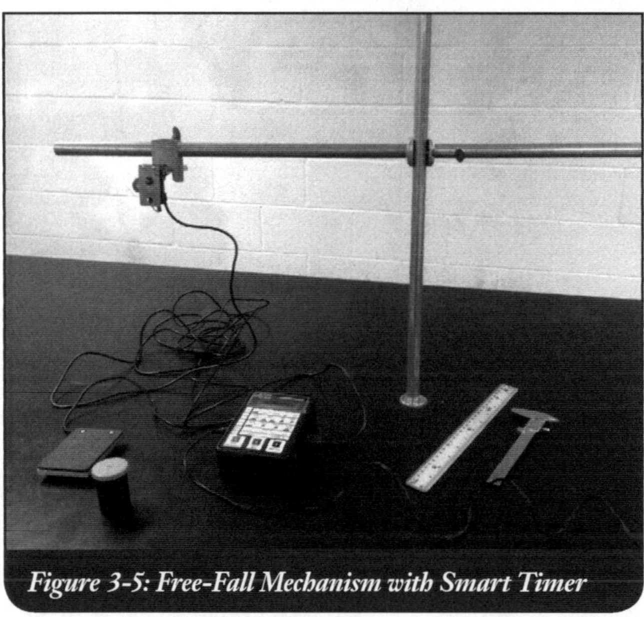

Figure 3-5: Free-Fall Mechanism with Smart Timer

If you are using the type of timer in Figure 3-1, press the RESET button just before releasing the ball. If you are using the Smart Timer, first press the red, SELECT MEASUREMENT button until the display reads TIME, then press the blue SELECT MODE button until STOPWATCH appears on the display, then press the black START/STOP button until you see an asterisk on the screen. You will need to reset the timer (either type) before each measurement. For the first type of timer, this involves pressing the small red button on the timer. For the second type, it involves pressing the black START/STOP button until an asterisk appears. You may also need to tap the pad before each drop as part of the reset process. Try this out a few times to become familiar with the procedure before you start to take measurements.

Once the timer is ready, release the ball by pulling very gently in a horizontal direction on the flat spring. Record the time of fall and the distance ("y" in Figure 3-4, measured with a ruler or meter stick graduated in millimeters; y is the distance from the bottom of the ball to the top surface of the target pad). Repeat the time measurement at least five times for a given distance. Use the average value in constructing your graphs. Don't forget to reset the timer before each measurement.

NOTE!

The timer may sometimes give spurious readings, particularly if you're using the first type described above. If a reading is clearly not compatible with other readings taken at that distance, disregard that reading and repeat the measurement.

P5. Before changing the distance y, repeat the five measurements using the ball of different size. Use the vernier caliper to measure the diameters of the two balls, and record the values on your data sheet. Also measure and record the masses of the balls. Note any differences, if any, between results using balls of different size.

P6. Repeat steps P4 and P5 for at least ten different values of y, ranging from about 10 cm to about two meters. To obtain the longer distances you will have to suspend the release mechanism out over the edge of the table and place the target pad on the floor.

CALCULATIONS

C1. Construct a table to display distance of fall (y) and time of fall (t). Be sure to include **all** the experimental data (Record all five trials at each distance, then calculate the average time for each distance. Use more than one table if you wish.)

C2. Construct a plot of y vs. t (using the average value of t for each value of y) and note the curvature.

C3. Construct a graph of log y vs. log t and note whether the plot appears to be linear.

C4. Use the method of least squares to determine the slope and the ordinate-intercept of the line of best fit, and superimpose this line on your plotted data. (You may use the linear fit option in Graphical Analysis or Excel, or use the method described at the end of Experiment 2.)

C5. Use the method of least squares to determine also the uncertainties in the slope and ordinate-intercept, and then report your values and associated uncertainties for the parameters k and n in Equation (1). Compare n with the theoretical value of 2 using a percent error, and compare 2 k with the value of g obtained in Experiment 2 using a percent comparison (if you have not yet done Experiment 2, omit this).

C6. Also compare your experimental value of g to the theoretical value of 9.8 m/s^2. (Use percent error.)

Kinematics in Two Dimensions: Projectiles

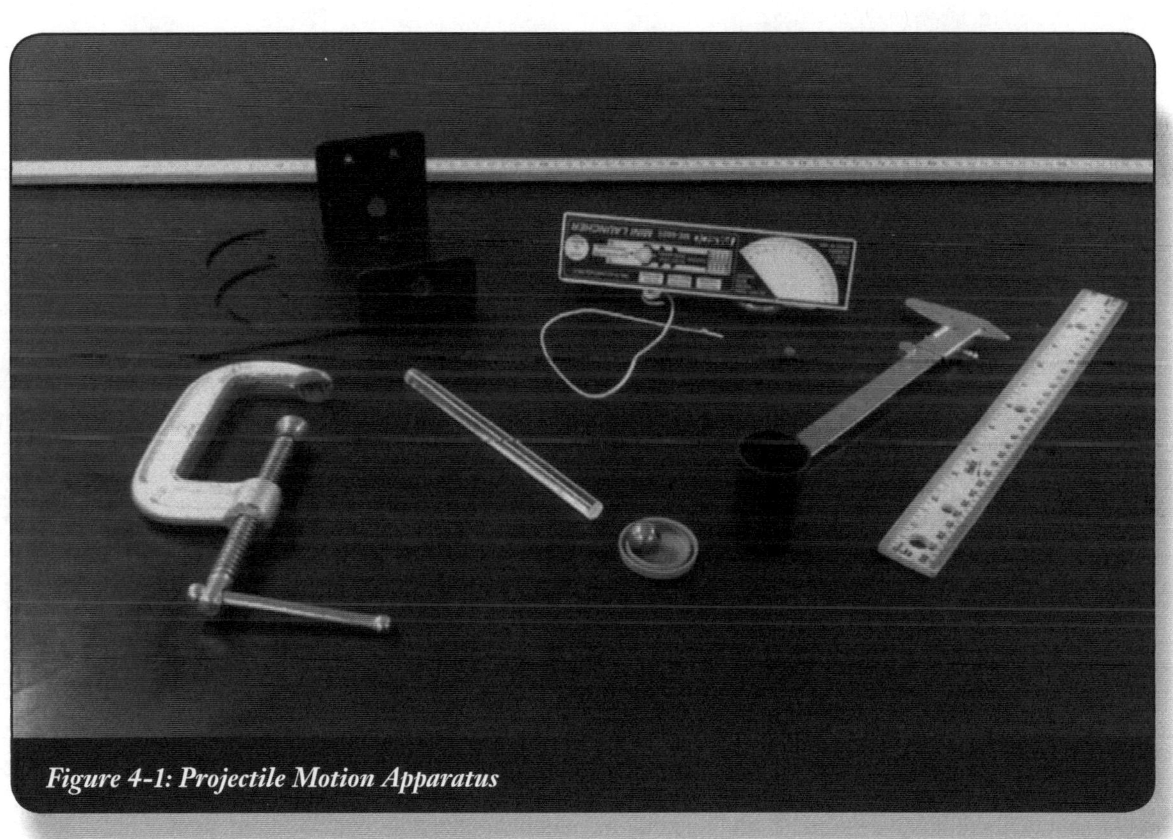

Figure 4-1: Projectile Motion Apparatus

INTRODUCTION

The kinematics of motion in two dimensions is studied here by using a spring-loaded "gun" to fire a small steel ball at varying angles from the horizontal. For an angle of zero degrees (the ball fired from a table surface above the floor) measurement of the horizontal distance of travel corresponding to the vertical distance of drop enables a determination of the initial speed of the ball. Measurements of the range acquired as a function of non-zero angles of projection then provide verification of the two-dimensional kinematical equations for projectile motion, including determination of the angle for maximum range.

PROCEDURE

P1. The apparatus consists of a spring-loaded firing mechanism with a ball which will serve as the projectile (Figure 4-1). Begin by securing the apparatus to the laboratory table with the clamp provided. Insert the ball into the firing mechanism and, by pushing inward on the ball, compress the spring until the ball locks into position. There are actually three different positions, or degrees of spring compression, at which the ball will lock. After locating the first locking position, find the second by pushing the ball and compressing the spring farther until the second locking position is found. Similarly find the third position. It may be helpful to use the short length of tubing provided to compress the spring mechanism. Pulling upward on the string attached to the top of the firing mechanism will release the ball.

P2. After you have become familiar with the firing mechanism, lock the ball in the first, or least compressed, position and adjust the angle for horizontal projection. Clamp the mechanism at the end of the laboratory table and, making certain no one is in the likely path of the ball, test fire the projectile onto the floor. Take precautions to ensure that after the initial impact with the floor the ball does not bounce into someone; fire the ball in such a manner that it bounces toward a wall, and place cardboard or other protecting material against the wall. Note where on the floor the projectile lands.

CAUTION!

While the velocity of the projectile in this experiment is relatively small, injury could result if the projectile should strike someone. Before firing the projectile make certain no one is in its path.

P3. To begin taking data, place a piece of carbon paper over a sheet of white paper and fasten them to the floor so that they are centered approximately at the point where the ball landed during the test firing. (Figure 4-2) The impact of the ball will leave a mark on the white paper. Now fire the ball. Run five trials and measure the horizontal distance traveled for each trial (from the crosshair on the apparatus marked "launch point" to the landing spot marked on the paper). Measure also the vertical distance of fall (from the crosshair marked "launch point" to the floor). These data will be used to determine the speed of the ball as it leaves the firing mechanism.

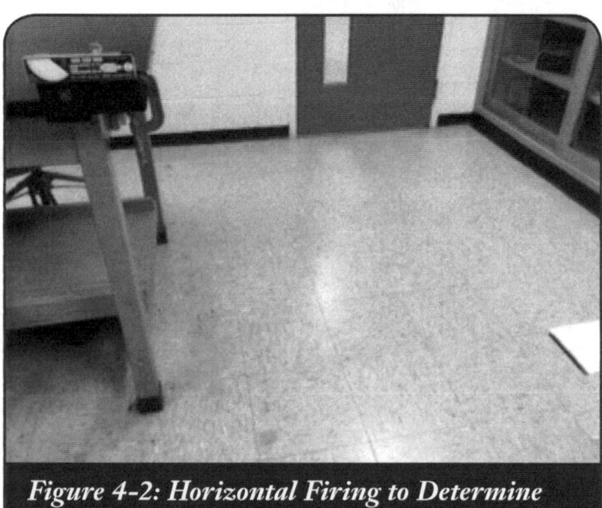

Figure 4-2: Horizontal Firing to Determine the Initial Velocity

P4. To investigate the effect of changing the angle of projection on the range, move the mechanism to the far end of the table so that the ball, rather than landing on the floor, will land on an the table at the same height as that from which it was fired (Figure 4-3). *Be certain that your projectile will not strike anyone.* Make a test firing, using an initial projection angle of 45°. This will yield a horizontal range between approximately 85 cm and 100 cm if the least-compressed trigger position is used. Once the proper positions have been established, place the carbon paper with the white paper underneath on the appropriate spot on the table and begin taking data: measure the horizontal distance traveled, or range, for a given angle of projection. Note that the protractor reads angles relative to the horizontal. Vary the angle from 10° to 80° in five-degree increments, taking at least three trials at each angle. The position of the paper will have to be re-adjusted every time the angle is changed (or several pieces of paper can be connected to cover the entire length of the table before beginning, as shown in Figure 4-3).

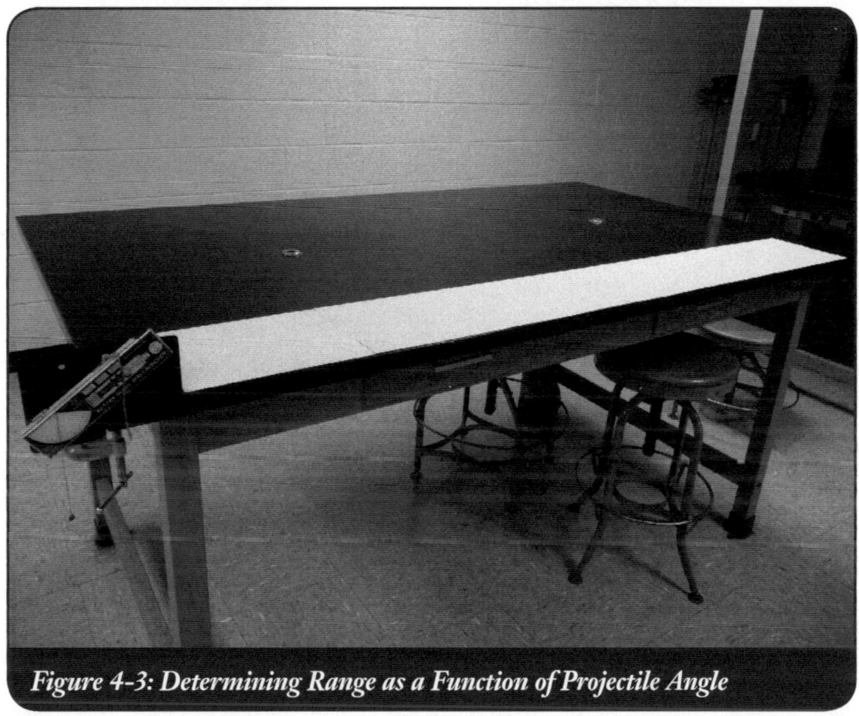

Figure 4-3: Determining Range as a Function of Projectile Angle

P5. If time permits, repeat steps P2–P4 using a more compressed position of the firing mechanism. This will provide data corresponding to a greater initial projectile speed. (You made need to be creative in providing a way to mark the landing spot since the ball will travel farther than one table length for some angles at these spring settings.)

CALCULATIONS

C1. When the ball is projected horizontally at the edge of the laboratory table as in step P3, the independent equations describing the horizontal (x) and vertical (y) components of the motion take on especially simple form. For the horizontal motion,

$$x = v_i t \qquad (1)$$

and for the vertical motion,

$$y = 1/2 \, g \, t^2 \qquad (2)$$

where v_i represents the initial speed of the projectile. Eliminating t between these equations and solving for v_i gives:

$$v_i = x \sqrt{\frac{g}{2y}} \qquad (3)$$

Using the data acquired in step P3, find v_i and its associated uncertainty.

C2. Using the data acquired in step P4, graphically plot the values of range R vs. projection angle θ. Note there is some uncertainty in *both* of these variables; the uncertainty in R can be found by averaging over the three trials taken at each angle, and the uncertainty in θ may be estimated. Show these uncertainties using error bars on your plotted points.

C3. Using the value of v_i obtained in C1, calculate the range for each projection angle, as predicted by the kinematical equation:

$$R = \frac{v_i^2 \sin(2\theta)}{g} \qquad (4)$$

Present your results in a table of four columns, the first for θ, the second for R calculated from Equation (4), the third for the experimental values of R, and the fourth for a percent comparison between the values in the second and third columns.

C4. Using Equation (4), carefully superimpose a curve showing the theoretical variation of R with θ onto the graphical plot of your data (step C2).

EXPERIMENT 5

Vectors: Forces and Static Equilibrium

Figure 5-1: The Three-Dimensional Force Apparatus

INTRODUCTION

In this experiment a plumb bob is suspended by three strings passing over pulleys; masses are attached to the free-hanging ends of the strings (see Figure 5-1). The tensions in the strings are adjusted by varying the values of the suspended masses and the positions of the pulleys until the plumb bob is in static equilibrium directly over the center of the table. Using the angular scale on the table and the vertical reference grid, the directions of the tensions are determined, and their magnitudes are found from the suspended masses. The condition for equilibrium is then verified by finding the vector resultant of the three tensions and the force of gravity acting on the plumb bob.

PROCEDURE

P1. Carefully examine the equipment. Use the small spirit level to level the table. Attach the three vertical risers, positioning them initially so there is no more than 160° between any adjacent pair.

P2. Pass the strings attached to the plumb bob over the pulleys and fasten a weight hanger to the free end of each string.

P3. Holding the plumb bob with one hand, add masses to the hangers until the plumb bob remains approximately over the center of the table when released. The apparatus should now resemble Figure 5-1.

P4. Now add or remove masses and adjust the angular positions of the vertical risers until the plumb bob hangs as precisely as possible over the center of the table.

P5. Record the angular position of each of the vertical risers on the horizontal disc. These are the θ coordinates of each mass. Also record the corresponding values of the suspended masses.

P6. Placing the protractor behind each string in turn, record the angle each string makes with the horizontal.

P7. Repeat steps P3–P6 for at least two more different combinations of masses and positions of the vertical risers.

P8. Determine the mass of the plumb bob using the balance provided.

CALCULATIONS

C1. For each configuration of vertical risers and masses, use the equations in Figure 5-2 to calculate the x, y and z components of each of the four forces acting on the plumb bob. Record your results in tables, with a separate table for each configuration. The forces due to the weights on each weight hanger may have components in each of the three directions. The x and y components may be positive or negative. The z components of each of these forces will be positive. The fourth force is that due to the plumb bob. The x and y components of this force will be zero. The z component will be equal in magnitude to the weight of the plumb bob and it will be negative. Be sure to include the signs of each of the force components in your tables. (Since all forces will be obtained by multiplying the measured masses by g, you may want to leave out that factor and simply use the masses. Check with your instructor on this.)

C2. Using the column containing the x components of each of the forces for your first configuration, sum the positive values and the negative values separately. Since the net force in the x direction should be zero if the system is balanced, the sum of the positive values should be equal to the sum of the negative values. So the percent difference between these values gives a measure of the error in this experiment. Calculate this percent difference by taking the difference between the positive sum and the negative sum and dividing by the average of these two sums, and multiplying by 100.

C3. Repeat C2 for the y and z components. Then repeat this entire process for the other two configurations.

C4. Compare the percent difference values obtained above to the uncertainties for the corresponding quantities. (Note: $\delta(\sin \theta) = \cos \theta \, \delta\theta$ and $\delta(\cos \theta) = \sin \theta \, \delta\theta$.)

Figure 5-2 below shows a right-hand Cartesian system, with the origin at the center of the table, the x-y plane parallel to the plane of the table, the x-axis passing directly above the 0° mark on the table, the y-axis directly above the 90° mark, and the z-axis vertically upward. (The values of T, θ, and ϕ that you have obtained are the spherical coordinates of the force vectors.)

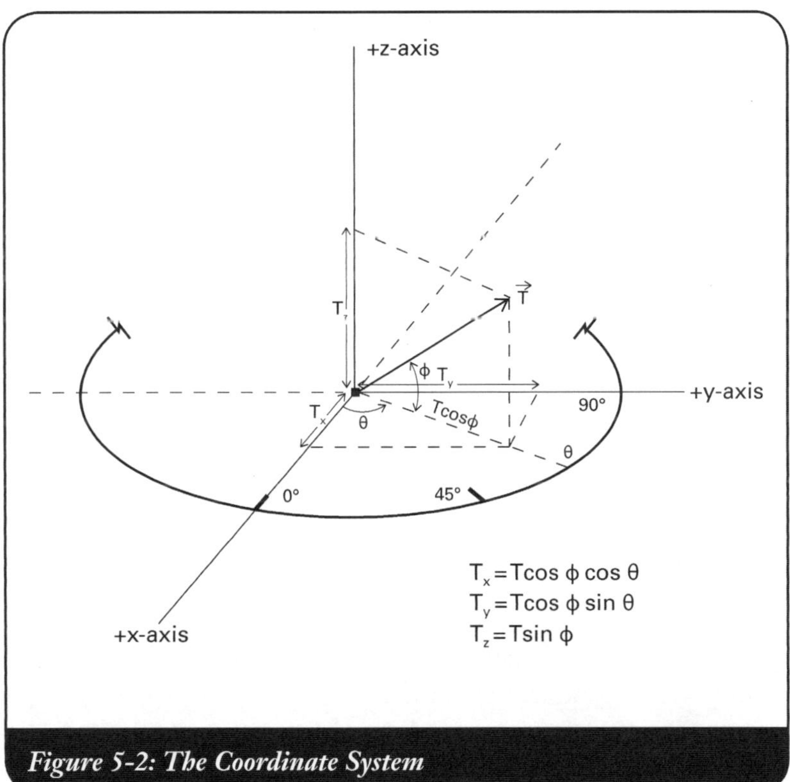

Figure 5-2: The Coordinate System

Notes

EXPERIMENT 

Statics, Dynamics, and the Force of Friction

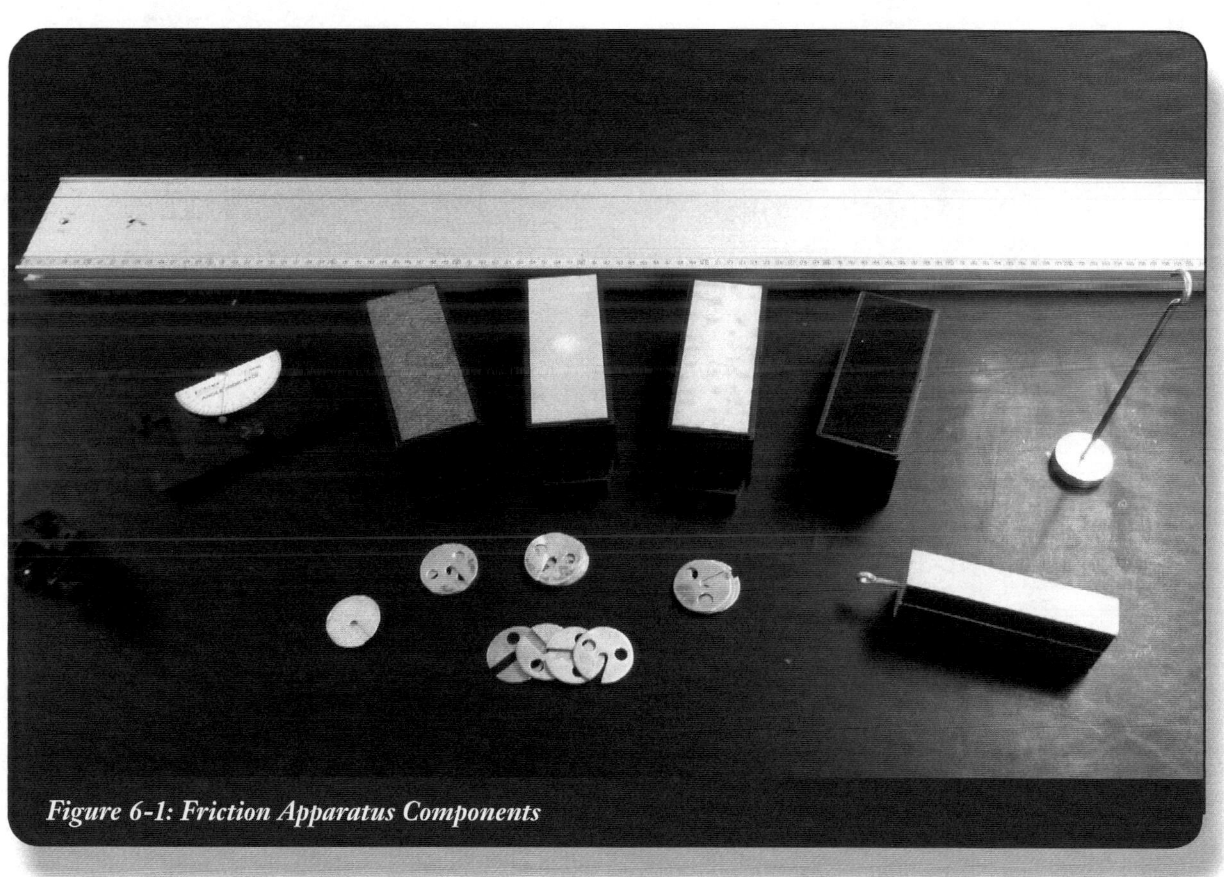

Figure 6-1: Friction Apparatus Components

INTRODUCTION

The coefficients of static friction between various surfaces are determined in this experiment by two different methods: first by finding the amount of force needed to cause a block to slide across a flat surface, and second by finding the angle at which a block begins to slip down an incline.

PROCEDURE

P1. Use the laboratory balance to determine the mass of each of the weight carts provided. Notice that there are three different types of surfaces on the bottom of the weight carts: felt, cork, and plastic.

P2. Begin with the apparatus assembled as shown in Figure 6-2. Choose one of the weight carts, attach a string to the front of the cart, and place the cart on the metal track. Attach the pulley to the end of the track and pass the string over the pulley. Attach a weight hanger to the free-hanging end of the string. With just enough mass in the cart to keep it from moving when there is no mass on the weight hanger,

CAUTION!

Please catch the cart to prevent damage to the equipment.

carefully add mass to the weight hanger until the cart just starts to slip, and record this mass. This provides data for one method for the determination of the coefficient of static friction between two surfaces.

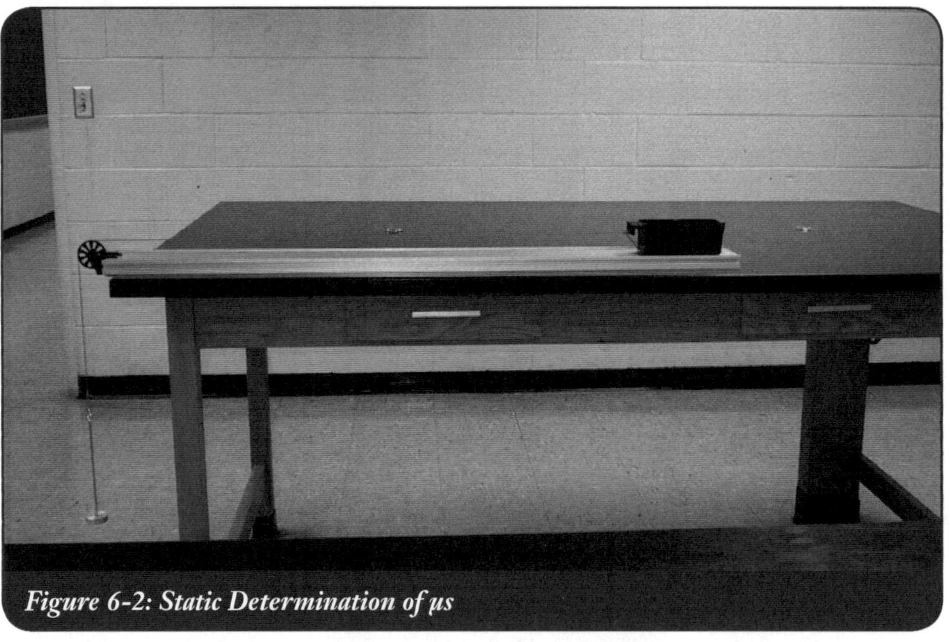

Figure 6-2: Static Determination of μs

P3. Repeat step P2 for the other two types of surface.

P4. Choose one of the surfaces and repeat step P2, but with additional mass added to the cart. Do this for at least five different masses, ranging from 100 grams to one kilogram of additional mass. This enables an investigation of the effect of the *normal* contact force on the force of friction.

P5. To investigate the effect of *contact area* on the force of friction, repeat step P2, but this time replace the weight cart with the block shown in Figure 6-3. By changing the orientation of the block, measurements can be made for two different surface areas for the felt surface and for two different surface areas for the wood surface, without changing the normal force. When finished, carefully measure the contact area for each orientation.

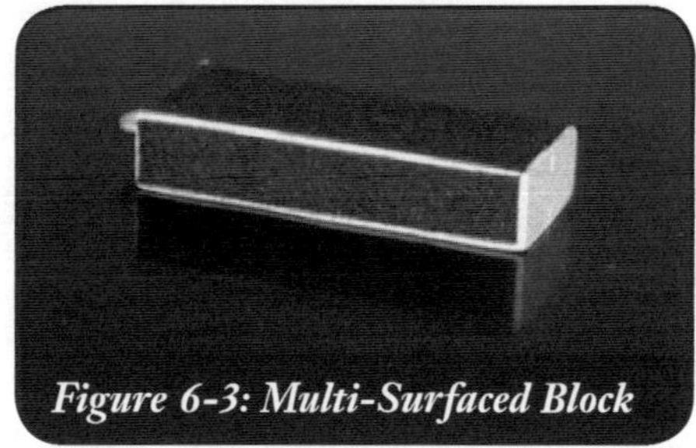

Figure 6-3: Multi-Surfaced Block

P6. Remove the weight hanger from the system. For each type of surface (plastic, cork, felt), slowly raise the incline until the cart just begins to slip (see Figure 6-4). Record the angle at which this occurs for each case. This provides data for a second determination of the static friction coefficients.

Figure 6-4: Incline Method of Determining μs

CALCULATIONS

C1. When, in step P2, mass added to the weight hanger is just sufficient to start the cart to slip, the tension in the string must be equal in magnitude to the maximum force of static friction acting on the cart. The tension is equal to the weight hanging at the end of the string (including the weight of the hanger). Use this information to show that the coefficient of static friction may be found by taking the ratio of the hanging weight to the weight of the cart. Calculate the coefficient of static friction for each type of surface (i.e., using the data acquired in steps P2 and P3).

C2. Use the data acquired in step P4 to determine what, if any, effects are produced on both the *force* of friction and on the *coefficient* of friction when the normal force is increased.

C3. Use the data of step P5 to determine the effects of varying the contact area. Again consider the effects on both the *force* of friction and on the *coefficient* of friction.

C4. A simple force analysis shows that when an object is just on the verge of slipping down an incline, the coefficient of static friction is given by the tangent of the angle of incline. Using the data acquired in step P6, calculate the coefficient of static friction for each type of surfaces studied, and compare your results with the calculations of C1. For each surface, find % difference between the results for the two different methods. Which method is more precise, and why?

Notes

Rotation and Newton's Second Law: Uniform Circular Motion

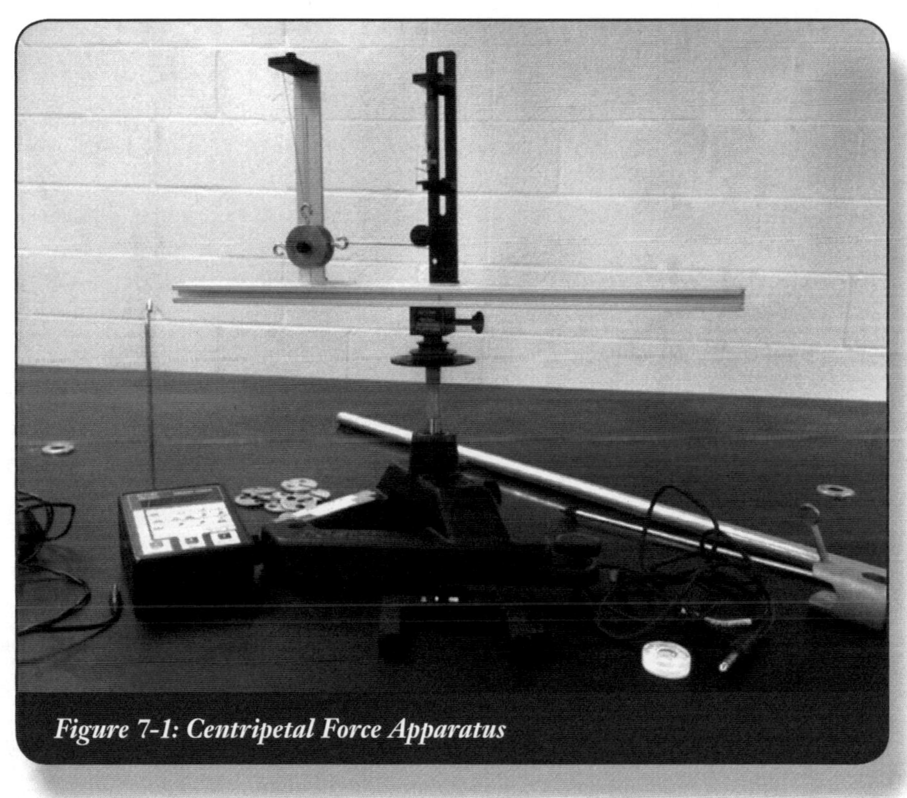

Figure 7-1: Centripetal Force Apparatus

INTRODUCTION

The centripetal force required to keep an object of mass **m** in uniform circular motion of radius **R** at speed **v** is given by

$$F = \frac{m\,v^2}{R} \tag{1}$$

If the period, **T,** of this motion is defined as the time required to complete one cycle, then

$$v = \frac{2\,\pi\,R}{T} \tag{2}$$

Combining Equations (1) and (2),

$$F = \frac{4\,\pi^2 m\,R}{T^2} \tag{3}$$

or

$$T = 2\pi \sqrt{\frac{m\,R}{F}} \tag{4}$$

The apparatus used here consists of a stand and a vertical shaft attached to a horizontal platform which can be rotated at a constant rate. Attached to the horizontal platform are two uprights. One of the uprights (the center post) is attached at the center of the horizontal platform and supports a spring and an indicator which allows the stretch of the spring to be maintained at a constant value. The other upright (the side post) is placed between the center and one end of the horizontal platform and supports a brass cylinder which is used as the rotating object. (See Figure 7-1.) The brass cylinder is comprised of three individual sections, two of which can be added or removed to adjust the mass of the cylinder. The side post can be moved along the horizontal platform to adjust the radius of the brass cylinder's motion. The spring on the center post and the brass cylinder on the side post are connected by means of a string that runs over a pulley at the bottom of the center post. Another string attaches to the opposite end of the brass cylinder, passes over another larger pulley placed at the end of the horizontal platform and attaches to a weight hanger that hangs vertically. (See Figure 7-2.) The amount of mass on the hanger regulates the force.

PROCEDURE

P1. Start by removing the brass cylinder, with all three sections included, from the side post and place it on the scale to measure its mass. Also measure its mass with one of the side sections removed and with both side sections removed. Replace the brass cylinder on the side post and use the small circular level to check the horizontal platform. Adjust the knobs at the bottom of each leg of the stand as necessary to level the horizontal platform.

P2. Check to make sure the position of the side post matches the radius label on the base of the apparatus. Adjust the side post as necessary, and record this radius. (Make sure the center post is positioned in the center of the platform.)

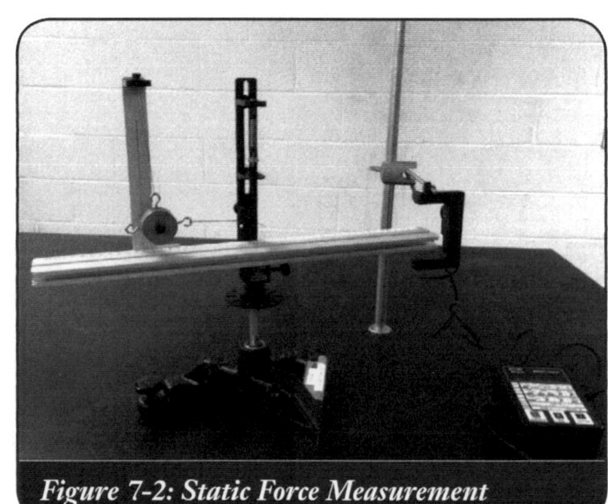

Figure 7-2: Static Force Measurement

P3. Using just the weight hanger with no additional weights (so that the hanging mass is just the mass of the hanger, i.e., 50 g), adjust the bracket that holds the spring on the center post until the brass cylinder hangs straight vertically." (Line up the string from which the brass cylinder is suspended with the black line on the side post to ensure that the cylinder is hanging vertically.) Adjust the bracket holding the brightly colored plastic indicator so that the indicator fits inside the bracket that holds the spring. (This will allow you to monitor the amount of stretch in the spring.)

P4. Now remove the pulley, the weight hanger and the string that attaches them to the brass cylinder and begin rotating the cylinder by spinning the vertical rod. Watch the plastic indicator and practice until you can maintain the indicator at a constant position inside the spring bracket while rotating the vertical shaft, which in turn rotates the horizontal platform.

ATTENTION!

Do NOT leave the weight hanger attached to the apparatus while it is being rotated.

P5. Once you are able to maintain a constant rotation you are ready to begin taking measurements. Position the photogate as shown in Figure 7-2 and plug it into the Smart Timer, using the side terminal marked "1." Press the leftmost button on the timer until it reads "TIME." Then press the middle button until it reads "PENDULUM." Then, while rotating the apparatus steadily, press the rightmost button to start the timer. You should hear three distinct beeps: the first when one end of the horizontal rail passes through the photogate, the second when the opposite end of the horizontal rail passes through the photogate, and the third when the first end passes through the photogate again. At the third beep, the timer stops and displays the time between the first and the third beep, which is the time for one complete cycle. If you do not hear three distinct beeps, discard the time and start over. Otherwise, record the time for that trial. Repeat this process three times and record the results for all three trials.

P6. The apparatus at each station in the lab will have the side post set at a different position. Once you have taken a set of measurements at the first radius, rotate to each of the other stations and repeat steps P2–P5. (Be sure to keep the amount of mass on the hanger and on the brass cylinder constant as you change stations. You may take your brass cylinder to each station with you if you wish.)

ATTENTION!

Do NOT change the radius on any apparatus except for minor adjustments to match the labeling on the base of the apparatus.

P7. Next remain at one station and repeat steps P3–P5, this time keeping the radius and the amount of mass on the brass cylinder constant, but adding mass to the weight hanger in 10 g increments until you can no longer raise the spring high enough to align the string holding the brass weight vertically. If you cannot get at least five different values for the amount of hanging mass (including the hanger), remove the mass hanger and use a paperclip (or just tie the masses on the end of the string) in order to get lower masses. Record data for at least five different masses, using three trials at each setting as before.

P8. Finally repeat steps P3–P5 keeping the radius and the amount of mass on the hanger constant but using the three different available possibilities for the mass of the brass cylinder (center portion only, center portion plus one extra section, and center portion plus both extra sections). Again record your data for three trials at each mass setting.

BE SURE TO REPLACE ALL PARTS OF THE BRASS CYLINDER AFTER COMPLETING THIS PART.

CALCULATIONS

C1. Take the average of the three time measurements for each setting and record that average as the period **T,** of the circular motion for each value of **R, m,** and **F.** (**R** refers to the position of the side post, **m** is the mass of the brass cylinder and **F** is the weight on the hanger, i.e., the mass on the hanger times g.) Make a separate table showing the variation of **T** with each of the variables **R, m,** and **F.**

C2. Using the values of **T** obtained in C1 for varying **R,** with m and **F** held constant, make a graph of log **T** vs. log **R.** Is the slope of this graph consistent with Equation (4)? Calculate % error and % uncertainty for the power dependence of **T** on **R.**

C3. Using the values of **T** obtained for varying **F,** with **R** and **m** held constant, graph log **T** vs. log **F** and again compare the slope of your graph with the result predicted by Equation (4). Calculate % error and % uncertainty as before.

C4. Finally, examine the dependence of **T** on **m.** Since you have only three different values for **m,** a graph is not useful in this case. Use Equation (4) to calculate the theoretical value of **T** for each of the values used for **m,** with **R** and **F** held constant. Compare each of these calculated values for **T** with the experimentally measured value. Compute % error and % uncertainty for each value of **T.**

EXPERIMENT 8

Inelastic Collisions and Momentum Conservation: The Ballistic Pendulum

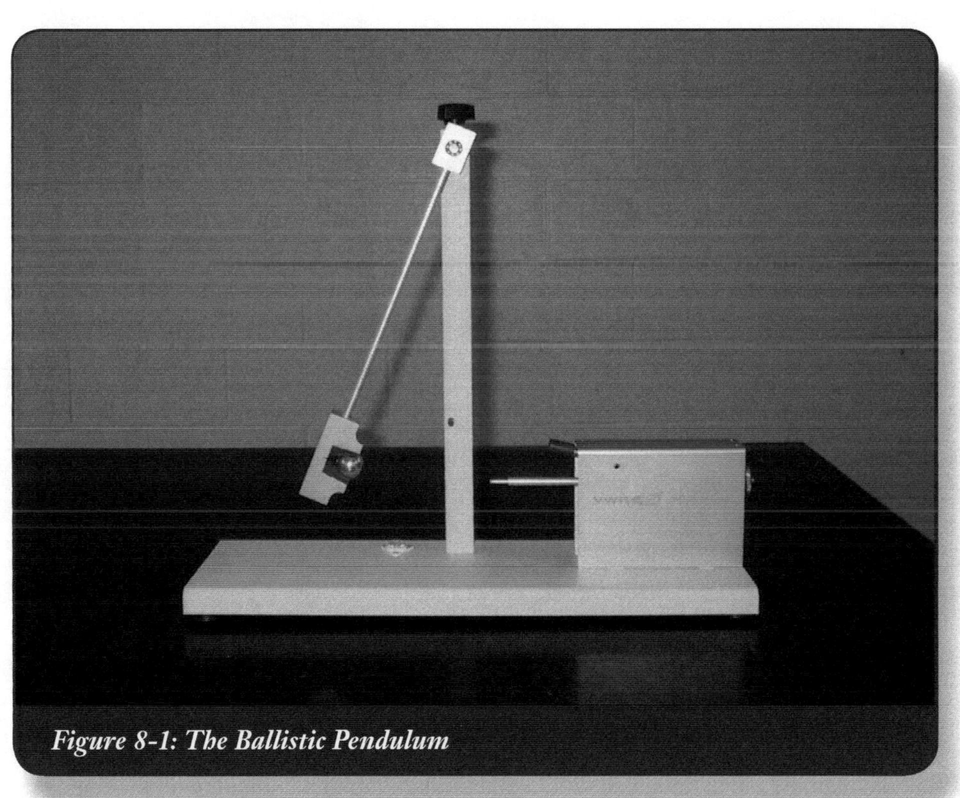

Figure 8-1: The Ballistic Pendulum

INTRODUCTION

As an application of the principles of energy and momentum conservation, the ballistic pendulum provides a method for determining the speed of a projectile. In the form of the apparatus used here, the projectile is a small steel ball fired from a spring-loaded "gun" similar to the one used in Experiment 4. After firing, the ball is caught in a completely inelastic collision by a freely-hanging pendulum in such a way that, after the collision, the pendulum-ball system swings upward. As the pendulum (with the ball embedded in it) swings upward, there is a mechanism that catches it just as it reaches its maximum height and would normally begin to swing back down, thereby holding the pendulum at the maximum height of its swing. The equation of momentum conservation for the inelastic collision, coupled with that for energy conservation as the pendulum-ball system swings upward, permits determination of the initial speed of the ball. As a check of the result, the pendulum is removed and the ball is fired horizontally from the edge of the laboratory table, the speed then being calculated from the horizontal distance traveled corresponding to the vertical drop of the ball, as in Experiment 4.

PROCEDURE

P1. There are two different types of apparatus used for this experiment. The physics is the same for both types. They differ only in the specific type of mechanism that catches and holds the pendulum-ball system after the collision. Begin by comparing your apparatus to Figures 8-1 and 8-2 to determine which apparatus you are using.

P2. If you are using the apparatus pictured in Figure 8-1, obtain the mass of the pendulum by reading it from the back of the apparatus. (See Figure 3.) If you are using the apparatus shown in Figure 8-2, detach the pendulum from the apparatus and measure its mass on the laboratory balance. Then reattach the pendulum to the apparatus.

Also measure the mass of the ball.

Figure 8-2: Alternate Ballistic Pendulum Apparatus

CAUTION!

Before firing the projectile, make certain no one is in its path.

P3. Measure the distance between the red dot on the pendulum and the flat surface of the base of the apparatus.

P4. Place the ball on the firing rod and push the ball back against the firing mechanism until the trigger engages and holds the ball against the compressed spring of the mechanism. On the first apparatus, the tension on the spring loaded mechanism can be adjusted continuously by turning the knob on the end opposite the rod that holds the ball. On the second apparatus, there are three positions of compression at which the trigger will engage. If you are using the second apparatus, you will need to take care to use the same spring setting for all measurements in this part of the experiment. (If you are using the first apparatus, this is accomplished simply by not turning the knob that adjusts the spring.)

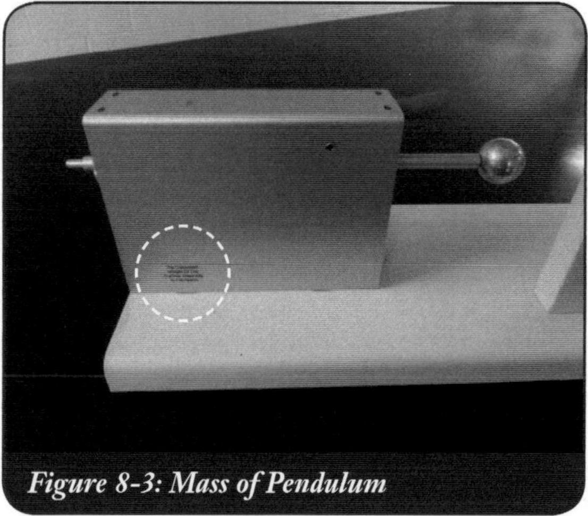

Figure 8-3: Mass of Pendulum

P5. Fire the gun by depressing the trigger pad. After the motion of the pendulum-ball system has been arrested at the maximum height, again measure the distance between the red dot on the pendulum and the base of the apparatus.

P6. Repeat the measurements of steps P4 and P5 at least four times, being sure to use the same initial compression position of the trigger mechanism each time.

P7. Place the apparatus near the edge of the laboratory table, and fire the ball onto the floor. You will need to move the pendulum out of the way. On the first apparatus this can be accomplished simply by loosening the knob, lifting the pendulum and retightening the knob. On the second apparatus, you will need to remove the pendulum. Again be sure to use the same initial compression that was used in steps P4 and P5. Using the technique described in Experiment 4, measure the horizontal and vertical distances traveled by the ball.

P8. Repeat steps P4–P7 for two other settings of the trigger compression. (On the first apparatus, you will need to turn the knob through several complete revolutions in order to make a noticeable change in the spring tension. On the second apparatus there are three different positions at which the ball will stop.)

CALCULATIONS

C1. If M represents the mass of the pendulum and m the mass of the ball, momentum conservation during the pendulum-ball collision implies a relation between the initial velocity v_i of the ball (i.e., before the collision) and the velocity V with which the pendulum-ball system begins its upward swing, i.e.,

$$mv_i = (m + M)\,V \qquad\qquad (1)$$

This in itself is not sufficient to determine v_i since, while the masses can be measured, V cannot be determined directly. After the collision occurs, however, the initial kinetic energy of the pendulum-ball system as the upward swing begins becomes converted to potential energy. The potential energy of the system at its highest point of swing must, in the absence of dissipative forces such as friction, be equal to the initial kinetic energy, i.e.,

$$1/2 \ (m + M) \ V^2 = (m + M) \ g \ h \qquad (2)$$

where h is the net increase in the height of the center of mass of the system during the swing. Using Equation (2) to eliminate V from Equation (1) and solving for v_i gives

$$v_i = \left(\frac{m + M}{m} \right) \sqrt{2gh} \qquad (3)$$

where g is the acceleration due to gravity. Note that since the energy and momentum conservation equations are ultimately based on Newton's second law of motion, 3) is a *dynamical* result. (See note that follows.)

Use the average of your measured values of m, M and h for a given compression of the trigger mechanism to find v_i. (Use the "book" value of 9.81 m/s² for g.) Remember, as always, to include a report of the uncertainty in your answer.

C2. Using the data from step P7 and the *kinematical* method of analysis given in Experiment 4, calculate v_i and its associated uncertainty.

C3. Compare the value of v_i obtained in step C1 above with that obtained in step C2. In your discussion, consider which method is more precise, and why.

C4. Repeat the analysis of steps C1–C3 for each of the other two trigger compression settings.

C5. Determine what fraction of the kinetic energy of the ball before collision is lost during the collision. What happens to it?

Note

The above analysis leading to Equation. (3) is, in fact, **not** accurate; the relevant corrections are discussed in P.D. Gupta, "Blackwood Pendulum Experiment and the Conservation of Linear Momentum," *Am. J. Phys.* **53** (3), 267 1985.

EXPERIMENT 9

Energy Conservation in Rotation

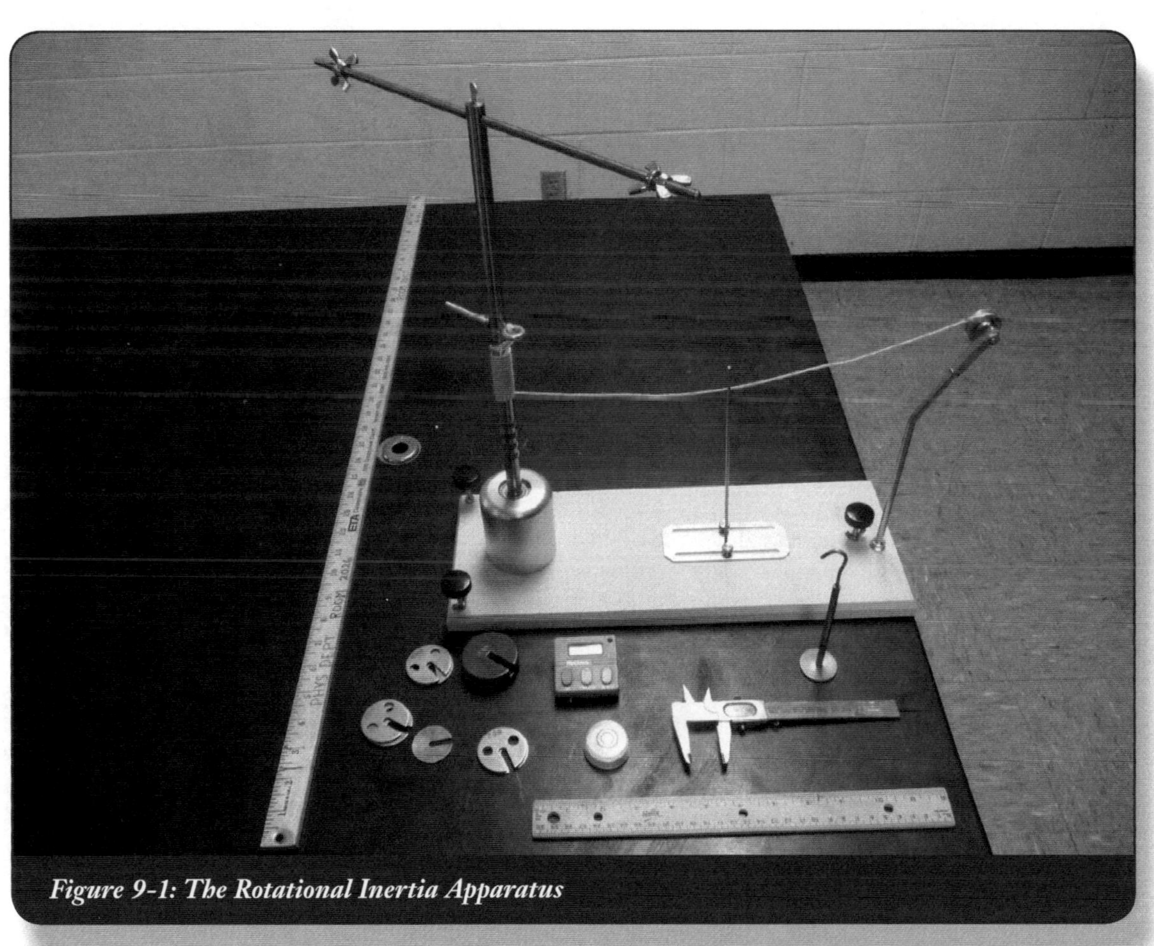
Figure 9-1: The Rotational Inertia Apparatus

INTRODUCTION

In this experiment a horizontal crossbar is set into rotation about a vertical shaft passing through its center. The crossbar is loaded with masses at given distances from the axis of rotation. The rotation is produced by a string wound around the axis and passed over a pulley, then attached to a suspended mass which is allowed to fall under the influence of gravity (see Figure 9-1). As the suspended mass falls, the resulting tension in the string produces a torque on the vertical shaft, giving a rotational acceleration to the crossbar and loaded masses.

Analyzed in terms of energy conservation, the decrease in potential energy of the falling mass is accompanied by a corresponding increase both in linear kinetic energy of the falling mass and in rotational kinetic energy of the crossbar, loaded masses, and vertical shaft. This principle is used to determine the moment of inertia of the loaded masses, and the result compared to the value predicted by theory.

PROCEDURE

P1. Carefully compare the equipment provided with that shown in Figure 9-1. Use the small spirit level to level the base of the apparatus. Measure the diameter of the vertical shaft with the vernier caliper. Be sure to include the string. Half of this diameter will give a measure of the distance from the center of the shaft to the center of the string, which is the radius (r) of the shaft needed for the calculations.

P2. Attach the crossbar to the vertical shaft, centering it as well as possible. Fix a 100-gram slotted mass to each side of the crossbar about 10 cm from the vertical shaft, using the wingnuts provided. Measure this distance carefully. This distance from the center of the vertical shaft to the rotating masses will be labeled R.

P3. Tie one end of a string about 130 cm long to the screw eye fixed to the vertical shaft. Pass the string over the pulley and fix the free end to a weight hanger. Add about 200 grams to the hanger and record this value as m. Rotate the vertical shaft so as to wind the string around the shaft and raise the hanger until the top of the hanger is near the pulley. Prevent the hanger from falling until you are ready to take data.

P4. When you are ready to take data, release the hanger and at the same moment start the timer. When the hanger reaches the end of its fall, stop the timer. Record the time of fall as t. Use a meter stick to determine the distance of fall (h).

P5. Repeat P4 several times to obtain good average values of time and distance.

P6. Remove the masses from the crossbar, *but leave the wingnuts in place.* Repeat the measurements of P4 and P5.

P7. For at least five different masses in turn, increase the mass (M) added to the crossbar from its initial value of 100 grams. Do not alter the distance (R) of the masses from the axis of rotation. For each choice of mass, repeat steps P4–P5.

P8. Return the masses on the crossbar to their original values of 100 grams, but now change the distance R from the vertical shaft. Repeat steps P4–P6 for at least five different values of this distance. Note that P6 must be repeated each time R is changed, whereas it did not need to be repeated each time the loaded mass, M was changed in P7; *why?*

CALCULATIONS

C1. As the suspended mass m falls through a distance h, its loss in potential energy mgh is accompanied by an increase in its kinetic energy, $mv^2/2$, and by an increase in the kinetic energy $I\omega^2/2$ of all rotating components. The angular velocity of rotation ω is related to the linear speed of the falling mass through the radius r of the vertical shaft, i.e., $v = r\,\omega$. The rotating components include the crossbar, loaded masses, wingnuts, vertical shaft, and pulley, as well as components not immediately apparent, such as bearings in the pulley and at the base of the shaft. Since the acceleration of the system is constant, the velocity of the falling mass at the instant the mass has fallen the distance h is just twice the average velocity during the fall, and this average velocity can be found by dividing the distance of fall by the time duration t of the fall. Using this reasoning, *show in your report* that I can be calculated from the relation

$$I = mr^2[(gt^2/2h) - 1]$$

From the data obtained in steps P4–P7, obtain values for the moment of inertia of the *loaded masses only* for each choice of loaded mass M you made in step P7. Do this by first calculating I for the loaded system (steps P4 and P5), then for the unloaded system (step P6), and subtract. Show your results in tabulated form and graphically, i.e., plot the moment of inertia of the loaded masses on the ordinate axis and the values of the loaded mass on the abcissa. How does I vary with mass M?

C2. Using the data obtained in P8, repeat the calculations of C1 for each value R. Again show your results graphically, this time plotting the moment of inertia on the ordinate axis as before, but on the abscissa plot the axis-mass distance, R.

C3. To the extent that the loaded masses can be considered point masses, the moment of inertia of the loaded masses should be given by $2MR^2$, M being the value of either of the masses (assumed equal) and R the distance from the axis of rotation to either mass (assumed the same for both masses). Tabulate your experimental values of I and compare to the theoretical value of $2MR^2$ for each value of M and R. Thus the graph obtained in C1 should be linear, but the graph obtained in step C2 should not. To further investigate the relation between I and R, do one of the following:

C4. a. Plot I vs. R^2, using least squares.

    b. Plot ln I vs. ln R to convince yourself the relation between I and R is of the power law form, then use a least squares fit to determine the exponent of R. Compare this to the theoretical value.

Notes

Simple Harmonic Motion: The Oscillating Mass-Spring System

Figure 10-1: The Mass-Spring System

INTRODUCTION

This experiment, while simple to perform, provides an excellent example of empirical deduction. A mass m, suspended from the lower end of a spring with fixed upper end, is set into oscillation. The fundamental observation is made that the motion of the mass tends to repeat itself in cycles, i.e., the mass oscillates up and down, each successive cycle an apparent close repetition of the previous cycle. Changing the mass reveals that the time for one cycle (the *period* of the motion) also changes, so it becomes of interest to determine the quantitative relation between mass and period. Such knowledge then permits the motion of any similar system to be predicted.

The theory of an idealized version of this motion is well known. Hooke's law $F = -kx$, when used for F in Newton's second law of motion $F = ma$, results in a mathematical equation which can be solved for the position x of the mass as a function of time. The equation is a second order differential equation (since a is by definition d^2x/dt^2), so you probably haven't seen the mathematical methods for obtaining the solution, but the result is:

$$x = A \cos (\omega t + \varphi)$$

The constant ω represents the frequency of the motion expressed in radians per second, and is related to the period T through the expression

$$T = \frac{2\pi}{\omega}$$

The mathematical analysis further shows that the frequency, and hence the period, are determined by only two parameters: the mass m and the stiffness constant k appearing in Hooke's law. This relationship is

$$T = 2\pi \sqrt{\frac{m}{k}} \tag{1}$$

The analysis leading to Equation (1) makes two simplifying assumptions. First is that the motion is truly periodic, i.e., that the system would, if left undisturbed, continue its motion indefinitely. You will readily discover that for the system used in this laboratory this assumption is not valid: the motion will noticeably decrease in amplitude after only a few cycles, principally due to the resistance offered to the moving mass by the air through which the mass must pass. The second assumption made in the mathematical analysis is that the spring itself contributes no inertia to the system, i.e., that the spring is massless. The larger the suspended mass relative to the mass of the spring, the less effect this assumption will have on your results. Unfortunately the spring used here is relatively massive, and its mass must be taken into account. The situation is complicated further by the fact that the spring used here is not of uniform cross-section, but rather is in the form of a truncated cone. This helps reduce interference between the windings of the spring and produces a motion which is closer to true simple harmonic motion than would be obtained with a simple helical spring. The corrections these complications necessitate in Equation (1) are derived using calculus in the Supplementary Discussion later, but the net result is to change the mass in Equation (1) from m, the mass of the suspended object, to very nearly (m + M/3), where M is the mass of the spring, i.e.,

$$T = 2\pi \sqrt{\frac{m + \dfrac{M}{3}}{k}} \tag{2}$$

While this relationship is given to you here, remember that in this experiment you are to take the view that the relationship is unknown, and that you are trying to develop it through *observation*. Note also that because you are given only one spring you will be unable to investigate the effects of changing the stiffness constant k; you will, however, have sufficient data to permit a verification of the dependence of T on k predicted by Equations (1) and (2).

PROCEDURE

P1. Use the laboratory balance to determine the mass of the spring.

P2. Suspend the spring as shown in Figure 10-2. The end of the spring with the smaller diameter should be clamped to the side of the table, using the clamp provided. The stiffness constant k of the spring is determined by aligning a meter stick parallel to the spring and recording the elongations produced by different masses suspended from the lower end of the spring. Begin with a 50-gram mass hanger, then increase the mass in 50-gram steps up to a total of at least 250 grams.

P3. With a given mass (say 100 grams) on the end of the spring, pull the mass down from its equilibrium position. Release it and at the same instant start the stopwatch. Pick a number of cycles, i.e., 30 or 40, and measure exactly the time it takes to complete that number of cycles. The period is then the total time of observation divided by the number of cycles observed. Repeat this observation at least three times to obtain an average value of the period.

Figure 10-2: The Suspended Spring

P4. Repeat step P3 for at least five different masses.

CALCULATIONS

C1. Using the data acquired in step P2, plot the values of the weights of the suspended masses vs. the corresponding elongations produced. If the weights are plotted on the ordinate axis, the slope of the resulting graph will equal numerically the stiffness constant k appearing in Hooke's law. Use the method of least squares to determine this value and its uncertainty.

C2. To obtain an empirical determination of the dependence of period T on mass m, begin by plotting on values of T vs. (m + M/3) (you may, at this point, assume the spring correction is M/3) obtained in steps P3–P5. You will note that the resulting plot is not linear, but rather curves downward away from a linear plot. This leads one to suspect a possible power law relation between T and (m + M/3), with a power of (m + M/3) less than unity (see Experiment 3), so the next step in determining this relation is to plot log T vs. log(m + M/3). If the resulting plot appears linear, determine the power of (m + M/3) from the resulting slope found from a least squares analysis. Also, the vertical intercept of this plot should numerically equal log $(2\pi/\sqrt{k})$, Compare the value of k determined this way with the value found in step C1.

C3. One additional graph will illustrate the effect of the mass of the spring. Suppose the assumption is made that Equation (1) should represent the results of your observations. Then if a plot is made with T^2 on the ordinate axis and m on the abscissa, the resulting graph should be linear with a slope equal to $4\pi^2/k$ and a vertical intercept of zero. Construct such a graph; you will find that the vertical intercept is non-zero. The reason is that the analysis leading to Equation (1) does not take into account the mass of the spring, i.e., a zero intercept implies that if there is no mass suspended there will be no simple harmonic motion and consequently no associated period of oscillation. A non-zero vertical intercept, on the other hand, would seem to imply that an oscillation exists even with no mass suspended; this of course corresponds to the oscillation expected from the mass of the spring itself. If the linear plot of T^2 vs. m is extrapolated into the second quadrant, the point corresponding to $T^2 = 0$ is represented by a negative mass value. This represents the mass that would have to be subtracted from the system to produce no oscillation. Equation (2) suggests that this should be approximately one-third the mass of the spring. Compare this value obtained from the graph with the value of the mass of the spring determined in step P1.

SUPPLEMENTARY DISCUSSION
THE EFFECT OF THE SPRING MASS

a. The Untapered Spring:

The spring (Figure 10-3) is considered to be composed of elements of infinitesimal length dx, each of infinitesimal mass dm. One end of the spring is fixed, the other oscillates with amplitude A. It is important to note that it is only the mass element at the free end of the spring which oscillates with amplitude A; all other elements oscillate with different amplitudes, each amplitude being dependent on how far the element in question lies from the fixed end of the spring. If this distance is represented by x, then the amplitude of oscillation y of this element is proportional to x, i.e., y = Cx, where C is a constant of proportionality. Clearly C = A/L where L is the total length of the spring. Note also that dm = (M/L)dx, where M is the total mass of the spring. The analysis proceeds by equating the total energy of all such oscillating elements to the energy some effective mass M' would have if it were oscillating with amplitude A. Since the energy of an oscillator of amplitude y is given by $ky^2/2$, or equivalently by $\omega^2my^2/2$,

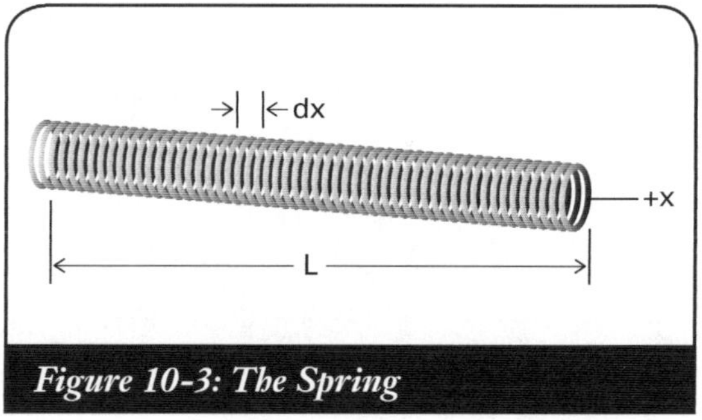

Figure 10-3: The Spring

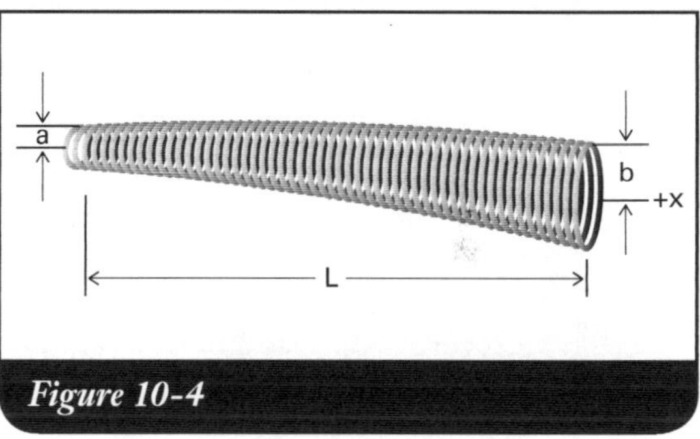

Figure 10-4

$$\int_{0}^{L} \frac{1}{2}\,\omega^2(dm)y^2 = \frac{1}{2}\,\omega^2 M'A^2$$

or

$$\int_{0}^{L} \frac{1}{2}\,\omega^2 \left(\frac{M}{L}dx\right)\left(\frac{A^2}{L^2}x^2\right) = \frac{1}{2}\,\omega^2\, M'A^2$$

or

$$\frac{M}{L^3}\int_{0}^{L} x^2 dx = M'$$

or

$$M' = \frac{1}{3}\,M$$

b. The Tapered (Truncated Cone) Spring:

Here a parameter must be introduced to characterize the taper. A convenient choice is $(b - a)/L$; call it τ (Figure 10-4).

If λ represents the mass per unit length of the wire from which the spring is wound, in one loop of radius r of the spring there is mass $m_1 = \lambda\,(2\pi r) = \lambda(2\pi)\,(a + \tau x)$. In a length dx (measured along the axis of the cone) there is mass $dm = (N/L)dx[\lambda 2\pi\,(a + \tau x)]$ (where N is the total number of loops in length L), and $dm = M$. Then, as in paragraph (a) above,

$$\int_0^L \frac{1}{2}\,\omega^2(dm)y^2 = \frac{1}{2}\,\omega^2 M'A^2$$

where $y = Cx = Ax/L$.

Thus,

$$\int_0^L \left(\frac{A^2}{L^2}\right)x^2\left(\frac{N}{L}dx\right)\lambda\,2\pi(a + \tau x) = M'A^2$$

or

$$M' = \frac{2\pi\lambda N}{L^3}\int_0^L (ax^2 + \tau x^3)dx$$

or

$$M' = 2\pi\lambda N\left(\frac{a}{3} + \frac{\tau}{4}L\right)$$

To eliminate λ and N, recall that

$$M = \int dm,$$

or

$$M = \frac{2\pi\lambda N}{L}\int_0^L (a + \tau x)dx$$

$$= 2\pi\lambda N\left(a + \frac{\tau}{2}L\right)$$

or

$$2\pi\lambda N = \frac{M}{\left(a + \frac{\tau}{2}L\right)}$$

Thus,

$$M' = M \left(\frac{\dfrac{a}{3} + \dfrac{\tau}{4}L}{a + \dfrac{\tau}{2}L} \right)$$

or

$$M' = \frac{M}{6} \left(\frac{4a + 3\tau L}{2a + \tau L} \right)$$

To check this result, note that for an untapered spring $\tau = 0$ so $M' = M/3$, the same result obtained in (a) above. For most of the springs used in the laboratory (but you should check this for your spring), $\tau = 0.031$ and $a = 1.5$ cm, so $M' = 0.34\,M$.

Notes

Wave Motion: Standing Waves and the Stretched String

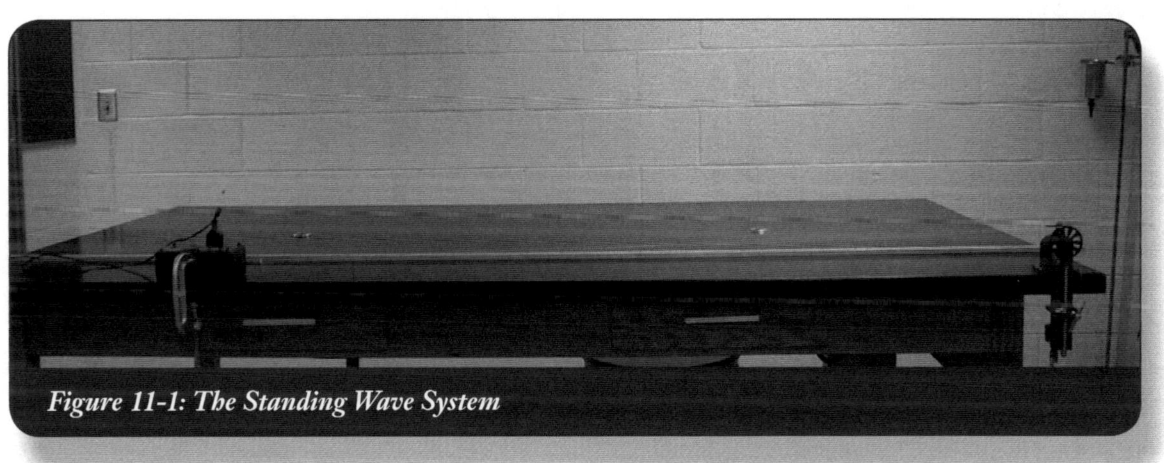

Figure 11-1: The Standing Wave System

INTRODUCTION

Two waves of equal amplitudes and frequencies traveling in opposite directions along a stretched string will produce a stationary, or standing, wave. If, as in this experiment, the ends of the string are fixed such that no oscillation can occur there, most of the energy carried by the waves is dissipated in damping at these points. Under certain circumstances, however, if the wavelength of the waves is just right, a wave traveling toward one of the fixed points will be reflected back in the opposite direction with little loss of energy. As a result most of the energy carried by the component waves goes into driving the string, and large amplitude vibrations of the string result. For this to occur, the wavelength must be such that the wave would tend naturally to have a node, or point of no amplitude, at the fixed point, so that no attempted driving of the fixed point occurs. This condition is satisfied if an integer number n of half-wavelengths fits evenly between the fixed points, i.e., if

$$n\frac{\lambda}{2} = L \qquad \text{or} \qquad n\lambda = 2L \qquad\qquad (1)$$

where λ is the wavelength of the component waves and L is the distance between the fixed points. The wave shown in Figure 11-2 is such that λ is ⅔ the length L, and three half-wavelengths fit between the fixed points. Each such pattern of loops with large amplitudes is a *mode* of vibration, the points where no vibration occurs (there are four of them in Figure 11-2, including the end points) are *nodes*, the positions of maximum vibration (halfway between the nodes) are *antinodes*, and the configuration of the string between adjacent nodes is a *loop*. The number of antinodes (or loops) is n in this form of standing wave.

The source of the waves used is an electrically-driven oscillator which produces a wave of frequency 60 hertz when the oscillator blade drives the string in a direction perpendicular to the length of the string (Figure 11-2); the frequency of the oscillator is not variable. (Note that, strictly speaking, the end of the string fixed to the oscillator is not at a node; while the amplitude of the vibration of the oscillator blade is small, it is not zero). The other end of the string is passed over a pulley, and tension in the string is provided by masses attached to this end.

The experiment begins with a simple observation: If the tension T in the string is varied, only certain values of T will produce modes of large amplitude, each value corresponding to a particular number of loops; the greater T, the smaller the number of loops, i.e., the longer the wavelength of the component waves. The object of the experiment then becomes to determine, through observation, the relation between T and λ. As in previous experiments, a power law is suspected, so the necessary graphs are generated to check this assumption, and the parameters obtained from the graphs can be interpreted and compared with the predictions of theory.

The theory for this relatively simple system is well known: Adjusting the tension in the string changes the velocity of the wave, since

$$v = \sqrt{\frac{T}{\mu}} \qquad\qquad (2)$$

where μ represents the linear density (i.e., mass per unit length) of the string. But changing the velocity produces a corresponding change in the wavelength, since the frequency f is fixed, and velocity, wavelength, and frequency are related by the fundamental expression

$$v = \lambda f \qquad\qquad (3)$$

Thus adjusting the tension produces large amplitude response (*resonance*) by obtaining just those wavelengths which provide an integer number of loops between the fixed points.

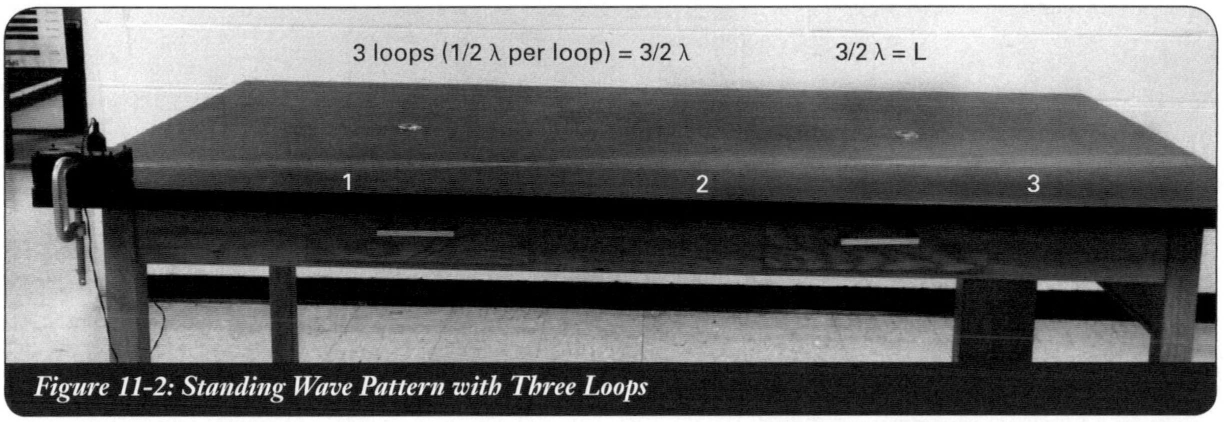

Figure 11-2: Standing Wave Pattern with Three Loops

PROCEDURE

P1. Obtain the value of μ (mass per unit length) for your string from your instructor.

P2. Establish the apparatus as shown in Figure 11-1; be certain the orientation of the oscillator relative to the length of the string is as shown in Figure 11-3.

P3. Place masses on the suspended hanger until a clear loop pattern is obtained; try to obtain at least seven or eight loops. This pattern is called a resonance **mode.** Fine adjustments may be made by moving the pulley toward or away from the oscillator. Once a good pattern is obtained, measure and record the total length of the string from the oscillator to the pulley. Also record the number of loops in the pattern and the amount of mass on the hanger. (Don't forget to include the mass of the hanger itself.)

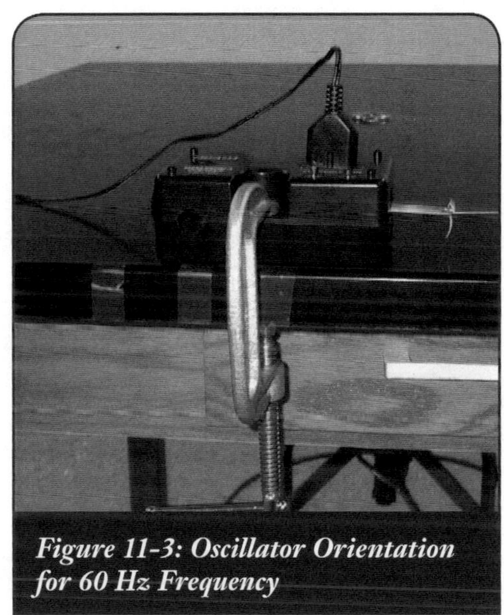

Figure 11-3: Oscillator Orientation for 60 Hz Frequency

P4. Increase the tension by placing additional mass on the hanger until another mode is obtained with fewer loops, and repeat the measurements of step P3. Continue this process until two or three loops are obtained, recording distances and corresponding tensions.

P5. Use the bracket and clamp provided to change the orientation of the ocillator so that the direction of oscillation of the string is parallel to the length of the string. (See Figure 11-4.) To get more than 4 loops, a mass of less than 50 g must be used (try a paper clip or a small mass). Repeat the measurements of steps P3 and P4.

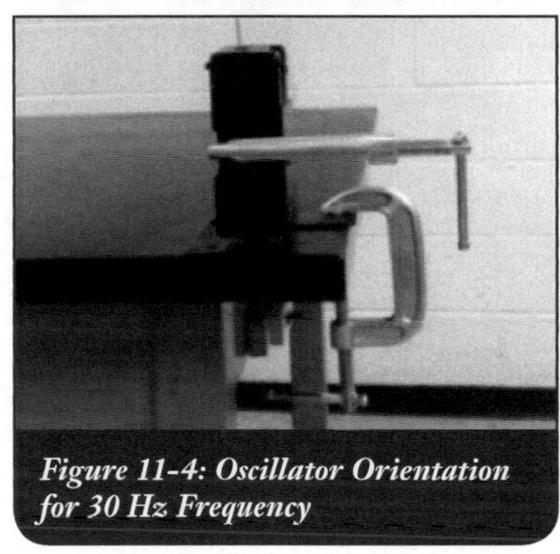

Figure 11-4: Oscillator Orientation for 30 Hz Frequency

CALCULATIONS

C1. Use the data from P3–P5 to calculate T for each resonance mode. (T is simply the weight placed on the string, which can be obtained by multiplying the mass on the hanger by g.)

C2. Use the data for the number of loops and the length of the string, along with Equation (1), to determine the wavelength for each resonance mode. Show all your results in tabulated form.

C3. From Equations (2) and (3) it is seen that theory predicts the relation between tension and wavelength, for fixed frequency and linear density, to be

$$T = \mu\, f^2 \lambda^2 \qquad\qquad (4)$$

Plot log T vs. log λ and obtain the line of best fit, determined by the method of least squares. Calculate the slope and its uncertainty for this line and compare your value to the theoretical value for that slope obtained by taking the ln of Equation (4). Calculate % error.

C4. Using the y-intercept from your log T vs. log λ graph and the value of μ from step C1, determine the frequency of the oscillator and the associated uncertainty; compare your result with the known value of 60 hertz. Calculate % error.

C5. Repeat steps C2–C4 using the data of step P5, but this time compare the frequency obtained to the expected value of 30 hertz. Again calculate the error and the uncertainty.

EXPERIMENT 12

Wave Motion: Standing Waves and the Speed of Sound and Beats

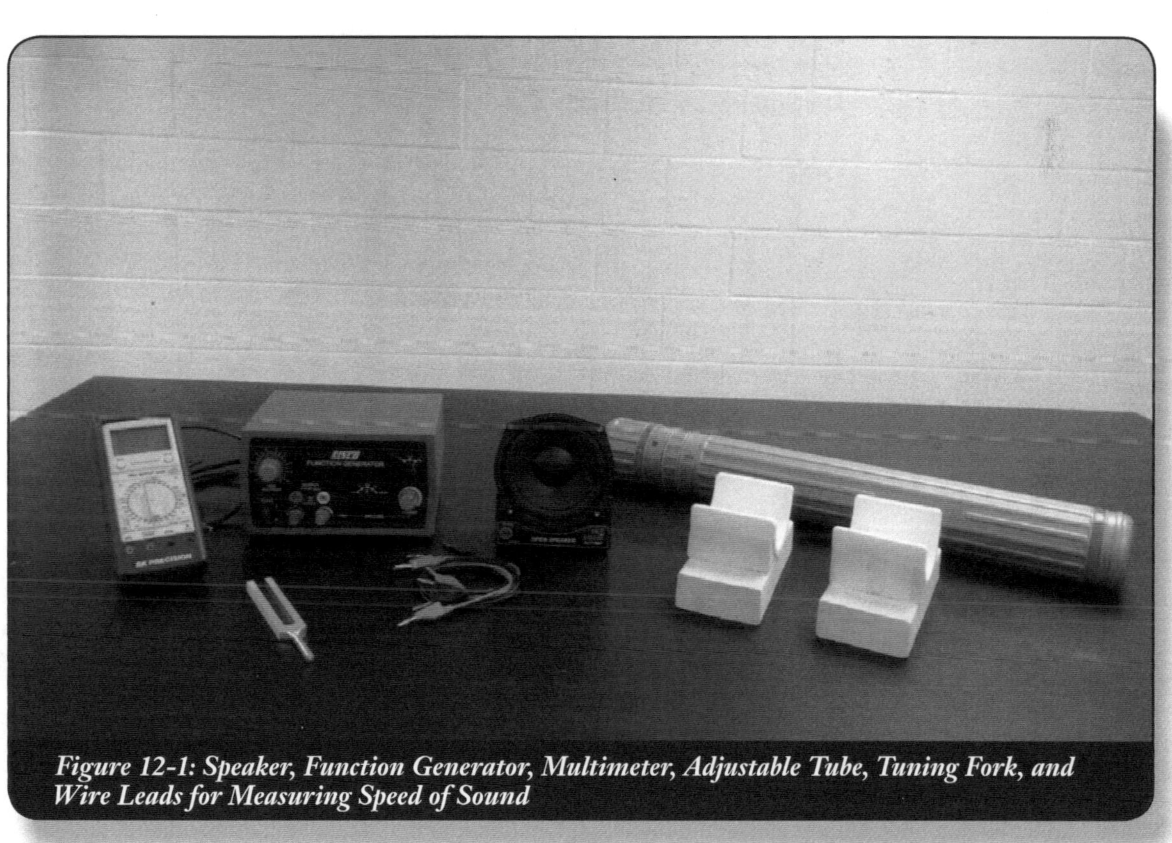

Figure 12-1: Speaker, Function Generator, Multimeter, Adjustable Tube, Tuning Fork, and Wire Leads for Measuring Speed of Sound

INTRODUCTION

In the vibrating string experiment, waves were generated in a string fixed at both ends, and a relationship was determined empirically between the tension T and the wavelength λ of standing waves at resonance. In that experiment the frequency of the source and the length and density of the string remained constant, and resonance was obtained by varying the wavelengths of the propagating waves. This was done by varying the tension in the string, which in turn varied the wave speed, which in turn varied the wavelength. When the wavelength was such that an integer number of half-waves exactly fit between the fixed ends of the string, resonance was detected through observation of the large amplitude response of the string. Coupling the observations with the predictions of theory enabled a determination of the frequency of the oscillator.

In the experiment you are about to do, resonance is also to be observed. But in this case the waves in question are the longitudinal disturbances produced by the passage of sound through air. The moving medium consists of the air molecules, whose motion cannot be observed visually; instead, resonance is detected by *listening* for the large amplitude response. Also, in this experiment, no property of the medium can be varied, i.e., there is no way to vary any property of the air analogous to the tension of the string in the vibrating string experiment. Sound here is produced by a function generator connected to a loudspeaker; the frequency of the function generator is variable, but for reasons discussed below you will take your measurements at a fixed frequency.

The question then becomes: if the frequency isn't to be varied, and no property of the medium can be altered to change the wavelength, how can the system be tuned to resonance? The answer lies in the way the air through which the waves pass is physically defined. In this case, the apparatus used is a pair of cylindrical tubes, one of which can be slid back and forth inside the other in order to adjust the length of the air column inside. Hence it is the *length* of the medium which is varied to tune the system to resonance. When, for a given frequency, the length of the air column is just right, the fixed wavelength will be such that interference between waves traveling down the column and those traveling up (after reflection from the other end) will produce a large amplitude standing wave, the presence of which is revealed by an increase in the intensity of the sound.[4] In the vibrating string experiment, the resonance condition occurred when an integer number of half-wavelengths fit between the fixed ends of the string, but in this experiment the boundary conditions are different. While there is an amplitude node at the closed end of the tube, the situation at the open end of the tube is somewhat more complicated. Sound waves consist of regions of compression and rarefaction of air molecules (analogous to the crests and troughs of transverse waves like those in the sring). While at the closed end, a compression is reflected as a compression and a rarefaction as a rarefaction, at the open end of the tube a compression is reflected as a rarefaction and vice versa; this occurs because a compression, when reaching the open end of the tube, rushes out, leaving a rarefaction to be reflected back into the tube.

Note

[4] It should be noted that the longitudinal sound waves can be described in terms of the variation in the amplitude of the vibrations (i.e., displacement of air molecules from their equilibrium positions), or in terms of the pressure variations in the air (or other medium). The pressure is at a maximum when the displacement is at a minimum, so that an amplitude node is a pressure antinode and an amplitude antinode is a pressure node.

The apparatus here allows for two different scenarios: the cap on one end of the tube can be left on so that one end of the tube is open and the other end is closed, or the cap can be removed so that both ends are open.

If the cap is left on, there is an amplitude node at the closed end and an amplitude antinode at the open end. As a result, resonance occurs when the shortest air column is one-quarter of a wavelength. (See Figure 12-2.) As the tube is lengthened, additional resonances occur whenever the length of the tube has increased a multiple of a half-wavelength beyond the position of the first resonance.

(Actually the situation is a bit more complicated, because the amplitude antinode at the open end of the tube occurs somewhat beyond the end, but that will not be important here since we will be using the difference in length from one resonance point to the next for our calculation.)

If the cap is removed, both ends are open, and thus there is an amplitude antinode at each end. In this case, resonance occurs when the shortest air column is one-half of a wavelength. (See Figure 12-3.) But as before, as the tube is lengthened, each successive resonance occurs as the length of the tube is increased by one half-wavelength beyond that of the previous resonance.

Once the wavelength which produces resonance for a given frequency is found, the analysis consists of determining the corresponding velocity of the sound wave and comparing with the value for the speed of sound predicted by theory.

In addition to the determination of the velocity of the sound, this experiment provides an apparatus, which enables you to investigate *beat* phenomena, i.e., the fluctuation in amplitude produced when two waves of slightly different frequencies interfere. A tuning fork of known frequency is sounded simultaneously with the signal from the function generator. The frequency of the function generator is measured with a multimeter, and the difference between the measured frequency and the frequency of the fork is compared with the observed beat frequency.

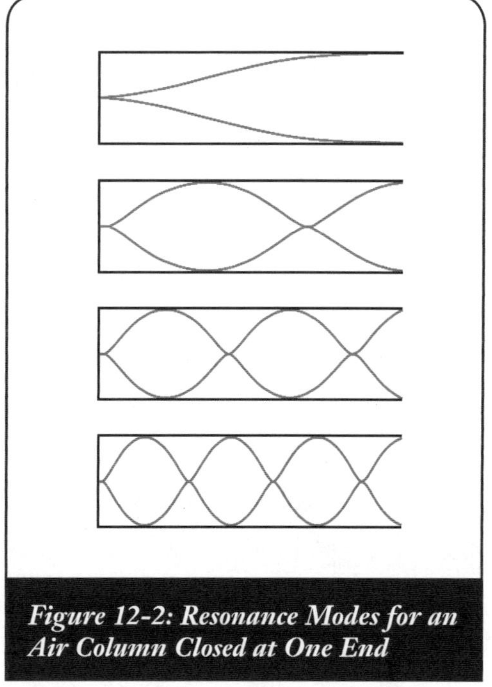

Figure 12-2: Resonance Modes for an Air Column Closed at One End

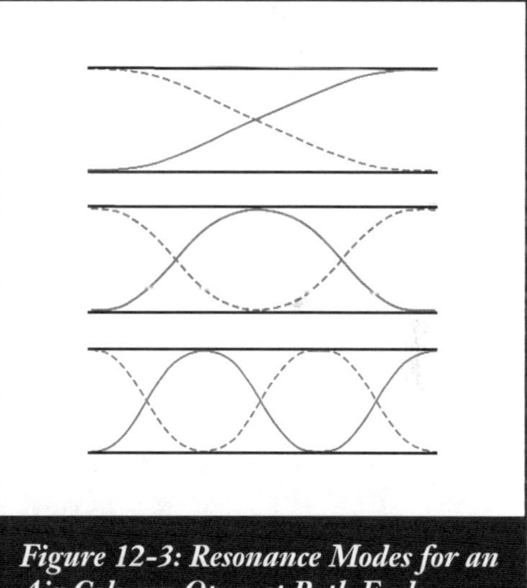

Figure 12-3: Resonance Modes for an Air Column Open at Both Ends

PROCEDURE

TUBE OPEN AT ONE END

P1. Connect the speaker to the function generator, using one set of the leads provided. Use the second set of leads to connect the **VΩHz** and **COM** terminals of the multimeter to the function generator and set the dial on kHz. (See Figure 12-4.) This will allow you to get an accurate measurement of the frequency of the function generator. Then place the adjustable tube (with the cap attached on one end) on the supports provided, and place the speaker about 2 cm from the open end of the tube. (See Figure 12-5.) Set the function generator to produce a frequency somewhere between 1000 Hz and 2000 Hz. (It's best if different groups do not use exactly the same frequency.) Starting with the inner tube pushed all the way into the outer tube, so that the air column inside is at its shortest length, gradually pull the inner tube out, thus lengthening the air column. Record the frequency from the multimeter and the length of the air column for each resonance, i.e., for each position where the sound is noticeably louder than it is on either side of that position. The length of the air column can be read directly from the tape that is affixed to the inner tube (Figure 12-6). (The length of the air column is actually several centimeters longer than the reading on the tape but, since it is only differences in length that matter, the reading from the tape for each resonance position is sufficient.) Repeat this process several times as you move the inner tube back and forth to make sure you have located all the resonance positions. (The interval between each pair of adjacent resonances should be approximately the same. If the gap between one pair seems larger than the rest, check to see if you missed one between them.)

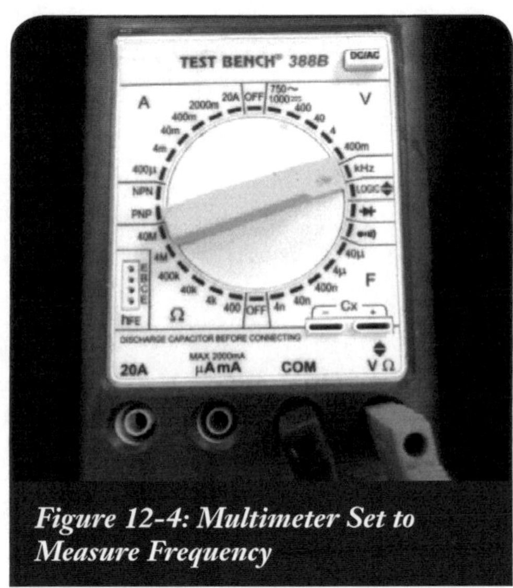

Figure 12-4: Multimeter Set to Measure Frequency

Figure 12-5: Setup for Finding Resonant Lengths

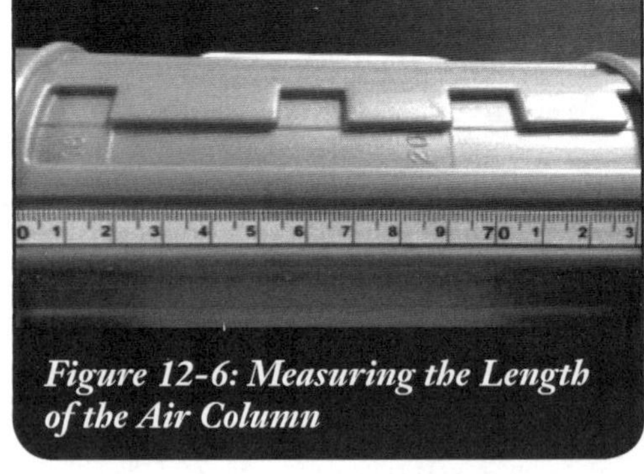

Figure 12-6: Measuring the Length of the Air Column

P2. Repeat step P1 using two frequencies higher than the one used in P1 but lower than 3000 Hz.

TUBE OPEN AT BOTH ENDS

P3. Remove the cap from the end of the tube and repeat steps P1 and P2, this time using three frequencies between 2000 Hz and 5000 Hz.

P4. The measurements taken in steps P1 to P3 will allow a determination of the speed of sound in air. Calculations from the data will give the wavelength which, when coupled with the measured frequency, permit the velocity to be found easily. These observations were taken at fixed frequencies. But the frequency of the oscillator is variable, so it would seem that a natural extension of the experiment would be to fix the length of the air column and vary the frequency, noting the frequencies which produce resonance. This turns out to be difficult, due to the nature of the psychological relationship between perceived loudness and actual sound intensity: perceived loudness is frequency dependent. For example, a signal directly from the speaker at 2000 Hz, with no resonance occurring, might appear just as loud as a signal of 150 Hz on resonance. For this reason it is difficult to judge the occurrences of resonances if the air column is held fixed in length and the frequency of the function generator is varied. Nevertheless, it is interesting to try. Only qualitative observations are required. Set the length of the air column to about 25 cm; since the speed of sound in the laboratory is approximately 344 m/s, a frequency of 344 Hz would produce a wavelength of one meter, and a column length of 25 cm would correspond to the first resonance.

Adjust the frequency of the oscillator to around 344 Hz until resonance is heard. Record the precise frequency from the multimeter. Now, **without changing the length of the tube,** calculate the next two positions or resonances expected and tune the function generator (without looking at the dial) to ascertain if they actually occur at these positions. (You may need to vary the tube length by a small amount to test whether resonance is occurring.)

P5. Read the thermometer in the laboratory and record the temperature.

P6. Select a tuning fork and note its frequency, stamped at the base. This value is accurate to 0.5%. Connect the speaker to the multimeter by plugging the two leads from the speaker into the terminals on the multimeter marked **VΩHZ** and **COM.** Then use another set of leads to connect these same two terminals of the meter to the blue and yellow terminals on the oscillator. Adjust the dial on the function generator to the range appropriate for your tuning fork. Adjust the frequency dial on the function generator to match approximately the value of the fork; the match should not be exact, but rather adjust this frequency so that a beat note between the fork, when struck, and the speaker is obtained. You want to determine the frequency of this beat note, so try for a frequency which will produce 15–20 beats in about 20 seconds, something that can be counted comfortably. Once a convenient beat frequency has been established, use a stopwatch to time 20–30 beats; the beat period is then this time interval divided by the number of beats observed, and the beat frequency is the reciprocal of this period. Record also the frequency of the function generator as measured by the multimeter. Note any discrepancy between the setting of the function generator dial and the reading of the meter; the meter is accurate to ±1 Hz. Repeat this process starting once with a frequency from the function generator that is higher than that of the tuning fork, and then one that is lower.

CALCULATIONS

C1. Using the data from steps P1 and P2, find the difference between each adjacent pair of resonance positions and record these distances as $\lambda/2$. Take the average of these values and multiply that average by 2 to determine the wavelength of the sound. Also calculate the uncertainty of this wavelength. (See the introductory pages of your lab manual or the tutorial on uncertainty if you don't remember how to do this.)

C2. From the wavelength obtained in C1 and the measured frequency, determine the speed of sound in the air column, using the equation

$$\lambda f = v \qquad\qquad (1)$$

C3. Repeat steps C1 and C2, but this time use the data from step P3.

C4. In the vibrating string experiment a theoretical equation was used to relate the physical properties of the string to the speed of the propagating waves, viz

$$v = \sqrt{\frac{T}{\mu}} \qquad\qquad (2)$$

A similar equation is obtained in the analysis of any mechanical wave: the wave speed is determined by two parameters, one characterizing the elasticity of the medium and one characterizing the inertia (T and μ, respectively, in Equation (2)). The form of this dependence is always the same:

$$v = \sqrt{\frac{\text{elastic factor}}{\text{inertial factor}}} \qquad\qquad (3)$$

In the case of a sound wave propagating in a gas, the explicit expression is

$$v = \sqrt{\frac{YP}{\rho}} \qquad\qquad (4)$$

where Y is a constant characteristic of the molecular nature of the gas, and for air is 1.40. P is the pressure of the gas, and ρ is the density. Contrary to intuition and the appearance of Equation (4), perhaps, the speed of sound is actually independent of pressure. This is because as the pressure increases, so does the density, if the temperature remains constant. This implies that the speed depends only on temperature, and the explicit temperature dependence can be found if a relation is known between pressure, temperature, and volume. To the extent that the air in the laboratory can be treated as an ideal gas, the ideal gas law, $PV = nRT$, is just such a relation. Using it to rewrite Equation (4),

$$v = \sqrt{\frac{YRT}{M}} \qquad\qquad (5)$$

Here R is the ideal gas constant (8.314 J/mol-K), T the temperature (in Kelvins), and M the molar mass of the air (0.0288 kg/mol).

Use the temperature from step P5 and the constants given above to calculate the theoretically predicted value of v from Equation (5), and compare with the values obtained in steps C1 to C3.

C5. Using the time recorded for a given number of beats, as observed in step P6, determine the beat frequency for both cases (one with the function generator set to a frequency higher than that of the tuning fork and one for a frequency lower than the tuning fork). Compare these values with the difference between the frequency of the tuning fork and the frequency read from the multimeter in each case. In addition, noting any discrepancy between the reading of the multimeter and the setting on the dial of the function generator, discuss the precision of the dial.

Notes

Notes

Notes

Notes

Notes

Notes